Graduate Texts in Contemporary Physics

Graduate Texts in Contemporary Physics

R. N. Mohapatra: **Unification and Supersymmetry: The Frontiers of Quark-Lepton Physics**

R. E. Prange and S. M. Girvin (eds.): **The Quantum Hall Effect**

M. Kaku: **Introduction to Superstrings**

J. W. Lynn (ed.): **High-Temperature Superconductivity**

H. V. Klapdor (ed.): **Neutrinos**

J. H. Hinken: **Superconductor Electronics** Fundamentals and Microwave Application

Johann H. Hinken

Superconductor Electronics

Fundamentals and Microwave Applications

With 94 Figures

Springer-Verlag Berlin Heidelberg New York
London Paris Tokyo Hong Kong

Professor Dr. Ing. Johann H. Hinken
Hans Kolbe & Co., FUBA Research Centre,
P.O. Box 1160, D-3202 Bad Salzdetfurth, Fed. Rep. of Germany

Translator: A. H. Armstrong
"Everglades", Brimpton Common, Reading, RG7 4RY Berks., UK

Title of the original German edition: *Supraleiter-Elektronik*
ISBN 3-540-18720-0 Springer-Verlag Berlin Heidelberg New York
ISBN 0-387-18720-0 Springer-Verlag New York Berlin Heidelberg

ISBN 3-540-51114-8 Springer-Verlag Berlin Heidelberg New York
ISBN 0-387-51114-8 Springer-Verlag New York Berlin Heidelberg

Library of Congress Cataloging-in-Publication Data. Hinken, Johann H., 1946-. [Supraleiter-Elektronik. English]. Superconductor electronics : fundamentals and microwave application / Johann H. Hinken : [translator, A.H. Armstrong]. p. cm. -- (Graduate texts in contemporary physics). Translation of: Supraleiter-Elektronik. Includes bibliographical references. (alk. paper : U.S.). 1. Superconductors. 2. Electronics. 3. Microwave devices. I. Title. II. Series. TK7872.S8H5613 1989 621.3--dc20 89-21577

Printed in the United States of America

2156/3150 - 543210 - Printed on acid-free paper

Preface to the English Edition

Recent research on superconductors with high critical temperature has led to results that were not available when the original German edition was prepared but could be included in the present English edition. This concerns materials based on bismuth and thallium, as well as measurements of low microwave loss.

The author would like to thank Mr. A. H. Armstrong for translating the book from German to English in a very dedicated and competent manner. Thanks are also due once again to Springer-Verlag for their generous support and cordial cooperation.

Bad Salzdetfurth
September 1989

Johann H. Hinken

Preface to the German Edition

The development of materials which lose their electrical resistance when cooled, even before reaching the boiling point of liquid nitrogen, has considerably increased the interest in superconductor technology, and with it superconductor electronics. This development had not been foreseen when work on the present book started, just over a year ago. Nevertheless, recent results of research on materials with high critical temperature are included to the extent that they seem to be confirmed and to be of interest to superconductor electronics.

The present book deals with the physical and technological foundations of superconductor electronics so far as they must be known in order to understand the principal modes of operation of superconductor electronics components. Special arrangements of such components can be advantageously employed, for example, in information technology, radio frequency technology, high precision electric and magnetic measurement technology, or electromedicine. Of these components those designed or used in microwave engineering will be considered in the text in more detail.

This book has evolved in part from notes for lectures intended for students of electrical engineering, especially radio frequency engineering, electronics and electrophysics at the Brunswick Technical University. The reader is expected only to be acquainted with the fundamentals of electronics and Maxwellian theory, and also, for Chap. 6, the elementary foundations of thermodynamics. The book should therefore be intelligible to all students of electrical engineering and physics. It is suitable not only as an accompanying text for lectures and self-study, and as an introduction to study in specialist fields, but also as a reference book for the practical engineer or physicist in research and development.

For critical review of individual chapters, thanks are due to Prof. Dr. K. H. Gundlach of the Institute for Millimetre Wave Radioastronomy in Grenoble, and to Dr. N. D. Kataria of the National Physical Laboratory in New Delhi. I am very grateful to Ms. A. Demmer and Mrs. B. Titze for their care in typing the manuscript and drawing the diagrams. I am also indebted to Mr. U. Klein and especially to Mr. R. Halx for frequent technical assistance, and to Springer-Verlag for their cordial cooperation during the preparation of this book. Finally I offer special thanks to my wife and my son for their forbearance when our family life all too often took second place.

Brunswick
November 1987 *Johann Hinken*

Contents

Introduction

Electronics is the special field of electrical engineering which is concerned with the effects of the movements of charge carriers in vacuo, in gases and in solids and which makes them useful in components and circuits for practical applications. The most important part of electronics is that of semiconductors. Much of our present mode of life is especially permeated with the transmission and processing of data and information. It is based on semiconducting materials, that is to say on those whose specific resistance lies between that of good conductors and good insulators.

Superconductors have an electrical resistance to direct currents (d. c.) which is immeasurably small. This effect arises at temperatures which are lower than the so-called critical temperature. Superconductor electronics is the sub-field of electronics which is concerned with the motion of charge carriers in and between superconductors. In the narrower sense one understands by superconductor electronics the electronics which is concerned with the applications of superconduction in communications technology or weak current technology, and not so much with its applications in electrical power engineering and high magnetic fields.

What do we now stand to gain from superconductor electronics which is new or better than that which is already available from the well-established semiconductor electronics? First of all, once the electrical resistance disappears so do the parasitic heat losses which occur in the circuit resistances and connections of semiconductor devices. The frequency limitations of superconductor components are therefore quite exceptionally high. If they are used as sensors their sensitivity limits can be so low as to be set only by Heisenberg's uncertainty principle.

Furthermore superconductor electronics can utilise special effects which do not occur with semiconductors. Above all these are the direct current (d. c.) and alternating current (a. c.) Josephson effects. They pertain to the quantum effects observable macroscopically. The SQUID magnetic field sensors, unsurpassed in their responsivity, have their operation based on the direct current Josephson effect. They make the quantisation of the magnetic field macroscopically observable and technically utilisable. With the alternating current Josephson effect the quantisation of electromagnetic field energies in the form of photons becomes macroscopically observable. This effect is put to technical use in high precision d. c. voltage standards.

That the electrical resistance vanishes in superconductors is true, strictly speaking, only for direct current. In a broad range of alternating current frequencies, however, the resistance is always many orders of magnitude lower than that of good normal conductors, even when these are also at low temperatures. The upper limit of frequency for the extremely low a. c. resistance of superconductors is presented by the circumstance that for any two electrons in the region of superconduction it is energetically advantageous, and therefore probable, that they together form a strongly bound pair – a Cooper pair. The electrons then become normally conducting again, only when enough energy is supplied to break the Cooper pairs. This energy 2Δ can be extracted from an alternating current, if its frequency is so high that the photon energy hf attains or exceeds the energy 2Δ. The superconductor behaves as a normal conductor only when the resulting frequency limit $2\Delta/h$ is exceeded. At temperatures which lie clearly below the critical temperature, this frequency limit amounts, for example, for Pb to about 650 GHz and for $YBa_2Cu_3O_7$ to about 7 THz. There are accordingly material-dependent upper frequency limits, below which only exceptionally small heat losses occur in superconducting electrodes and connections.

Upper frequency limits of the order of Δ/h also occur, however, for the functional mechanism of superconducting components which are based on the quantum mechanical tunnelling of single electrons and Cooper pairs. We shall see that the SIS mixer is an example of tunnelling by single electrons.

For Josephson junctions used as sensitive high frequency receivers the current carried by the tunnelling of Cooper pairs becomes smaller and smaller at high frequencies. At a characteristic frequency it becomes equal to the current of the normally conducting electrons which are also still present, flowing parallel to the Cooper pair current. At frequencies lying far above this characteristic frequency the Cooper pair current is much smaller than the resistive current. The special properties of the Josephson junction then disappear. For good quality Josephson junctions this characteristic frequency also lies in the order of magnitude of Δ/h.

The upper frequency limits of superconducting electronic components therefore already lie far outside the microwave region, which one normally associates with frequencies between 300 MHz and 300 GHz, corresponding to wavelengths of 1 m to 1 mm. Most high frequency technology applications of Josephson junctions so far, however, are in the microwave region. The possibility of introduction at higher frequencies is nevertheless always there. This is particularly so if materials with higher energy gaps 2Δ are used.

As we shall see, the energy gap 2Δ also limits the maximum d. c. voltage which can be represented in precision d. c. voltage standards to about Δ/e per Josephson junction. The energy gap 2Δ therefore turns out to be of quite critical significance for the limitations of superconductor electronics.

The development of materials with high critical temperatures and therefore, as we shall see, also high energy gaps 2Δ, accordingly extends the capabilities of superconductor electronics quite significantly.

Since wavelengths for propagation in free space and along transmission lines become shorter and shorter towards high frequencies, the finite size of components and their connections plays an ever more significant role. A compact circuit type in the form of an integrated circuit then becomes crucially important. Because of the small cross-sectional dimensions of the conductors the heat losses in semiconductor and normal conductor integrated circuits are of course quite significant. With superconducting integrated circuits, however, the losses are smaller by orders of magnitude. For example, Josephson d. c. voltage standards driven at 70 GHz using chips with 1447 Josephson junctions achieve a degree of integration much higher than that of monolithically integrated microwave semiconductor circuits at these frequencies.

1. Fundamentals of Superconductivity

Before we concern ourselves with superconducting components, the fundamentals of superconductivity will be treated in this chapter. We shall therefore describe the physical phenomena only in such detail as is necessary for an understanding of the SIS junction and Josephson junction in microwave circuits. More detailed presentations are to be found, for example, in [1.1–3].

1.1 Basic Phenomena

The remarkable phenomenon of superconductivity is the vanishing of the electrical resistance below a critical temperature T_c. The temperature dependence of the resistivity ϱ of normal conductors and superconductors is sketched in Fig. 1.1. At high temperatures thermal lattice vibrations determine the mean free path length of the conduction electrons and hence the resistivity ϱ. As the temperature decreases, so does ϱ. At low temperatures the thermal lattice vibrations no longer play a role in normal conductors. Imperfections in the otherwise periodic lattice potential determine the free path length. As $T \to 0$ the resistivity of normal conductors therefore tends towards a residual resistivity, which is smaller, the purer the normal conductor is.

Superconducting materials behave above the critical temperature T_c like normal conductors. When $T < T_c$, however, the resistivity is too small to measure. The transition from the normal conducting to the superconducting state takes place in a transition range which may be only a few mK. In an experiment by *Gallop* [1.4] the resistivity of a superconducting wire was found to be less than 10^{-24} Ωcm. This is a factor of 10^{-18} lower than the resistivity of copper at room temperature.

The critical temperature is known also as the transition temperature. It differs widely for different materials. Table 1.1 shows the transition temperature for a few useful superconductors. Moreover, ordinary solder is already superconducting at a temperature of 4.2 K. This is interesting, because many low temperature experiments are carried out at this temperature. 4.2 K is the boiling point of liquid helium at ambient pressure. A copper wire covered with solder is therefore superconducting in liquid helium.

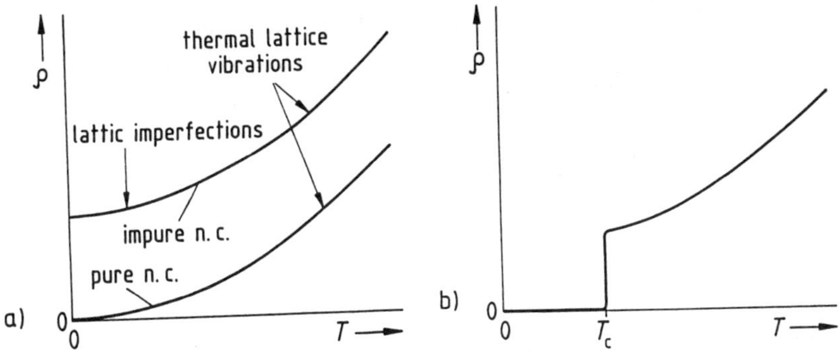

Fig. 1.1a, b. Qualitative behaviour of the resistivity ϱ as a function of the temperature T. **(a)** Normal conductor. **(b)** Superconductor

Table 1.1. Material parameters of a few superconductors

Material	Al	In	Sn	Pb	Nb	NbN	Nb_3Ge	$YBa_2Cu_3O_7$
T_c in K	1.2	3.4	3.7	7.2	9.2	$\simeq$ 16	23	92
$\lambda(T=0)$ in nm	50	64	51	39	85	$\simeq$ 200	$\simeq$ 150	140
2Δ in meV	0.34	1.05	1.15	2.7	3.1	$\simeq$ 4.8	7.8	$\simeq$ 30

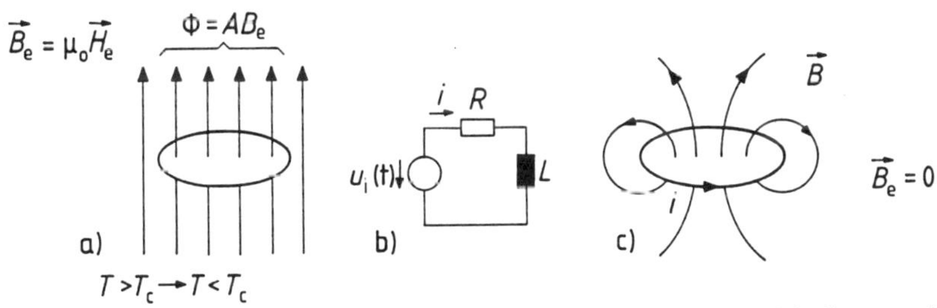

Fig. 1.2a-c. Closed resistance-free loop. **(a)** Cooling from $T > T_c$ to $T < T_c$ in the external magnetic field $\boldsymbol{H}_e$. **(b)** Equivalent circuit for the loop for current calculation as the magnetic field changes. **(c)** Distribution of the magnetic field when the external magnetic field is switched off.

One consequence of the vanishing electrical resistivity is that the magnetic flux through a closed resistance-free loop cannot vary. To understand this let us consider Fig. 1.2a. The loop is first of all at a temperature $T > T_c$. An external magnetic field with flux density $B_e = \mu_0 H_e$ is assumed to thread the loop perpendicularly with the magnetic flux $\Phi = AB_e$. The loop is cooled to $T < T_c$ whilst maintaining the external magnetic field B_e. Now the external magnetic field B_e changes. According to the Lenz law there is a contrary voltage induced in the loop:

$$u_i(t) = -A\frac{dB_e}{dt} \quad . \tag{1.1}$$

According to Fig. 1.2b this voltage source lies in series with the resistance R and the inductance L of the loop. Because the sum of all voltage drops in this loop must vanish (Kirchhoff's second law), we have

$$-A\frac{\mathrm{d}B_\mathrm{e}}{\mathrm{d}t} = Ri + L\frac{\mathrm{d}i}{\mathrm{d}t} \quad . \tag{1.2}$$

For the case under consideration here, where the resistance R vanishes, (1.2) is simplified to

$$-A\frac{\mathrm{d}B_\mathrm{e}}{\mathrm{d}t} = L\frac{\mathrm{d}i}{\mathrm{d}t} \quad . \tag{1.3}$$

Integration over time yields

$$Li + AB_\mathrm{e} = \text{const.} \tag{1.4}$$

The total flux $Li + AB_\mathrm{e}$ through the loop therefore remains constant. Although before the change in the external magnetic field the current i was zero, there is now a current. Whilst the total flux through the loop remains constant, the distribution of the flux density over the area A of the loop may vary. With the removal of the external magnetic field to $B_\mathrm{e} = 0$ a field distribution may be established in the neighbourhood of the loop as in Fig. 1.2c.

In addition to this phenomenon in a closed resistance-free loop, in a closed superconducting loop there occurs a quantisation of the magnetic flux. The total flux through a closed superconducting loop can accordingly assume only those values which are integral multiples of the flux quantum

$$\Phi_0 = \frac{h}{2e} \simeq 2.07 \times 10^{-15}\mathrm{Wb} = 2.07\mathrm{mVps} \quad , \tag{1.5}$$

see Sect. 1.5.

Nonresistive perpetual currents find application in superconducting magnets. Figure 1.3 shows a magnetic coil at a temperature below its critical temperature. With the circuit open a current is first of all sent through the coil from an external current source. When the circuit is closed the current goes on flowing as a perpetual current. The wires leading outside can now be switched off. The current and hence also the magnetic field of the coil remain constant.

The question now arises how a direct current is distributed in a parallel circuit of resistance-free branches. Let us consider Fig. 1.4. Because of the vanishing resistances in branches B and D Kirchhoff's second law (the Mesh rule) gives no result, as long as i is constant. We therefore consider the procedure by which the current i was built up. We can regard the switching on process as continuously changing the resistance R in Fig. 1.4 from $R = \infty$ to a finite value. During this timewise change of the total current i potential differences occur in the branches B and D because of their non-vanishing inductances. These potential differences must be equal

$$L_B\frac{\mathrm{d}i_B}{\mathrm{d}t} = L_D\frac{\mathrm{d}i_D}{\mathrm{d}t} \quad . \tag{1.6}$$

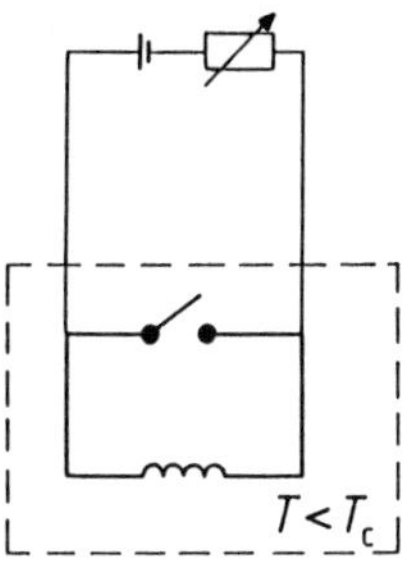

Fig. 1.3. Principle of the superconducting magnet

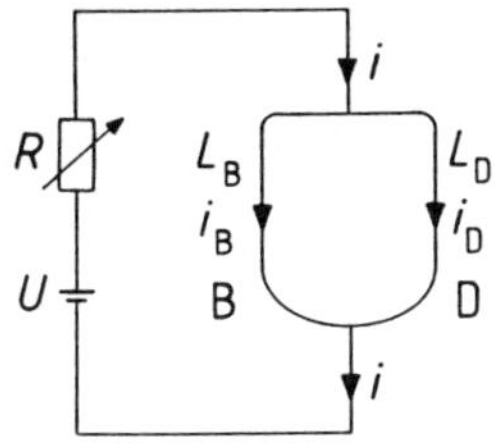

Fig. 1.4. Parallel resistance-free current branches B and D with inductances L_B and L_D

Here L_B and L_D are the inductances of the two branches. Integration of (1.6) leads to

$$L_B i_B = L_D i_D + \text{const} \quad . \tag{1.7}$$

Since at time $t = 0$ the current in both branches $i_B = i_D = 0$, the constant is also equal to zero. In the steady state, therefore, the ratio of the branch currents to one another is equal to the reciprocal of the ratio of the branch inductances. This reasoning can also be carried through in a similar way if their mutual inductances are taken into account.

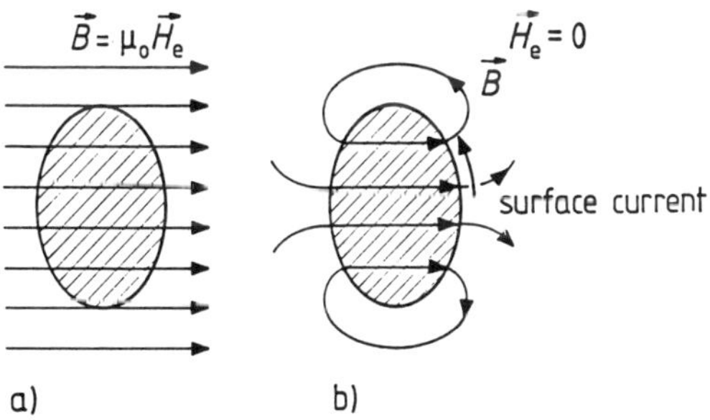

Fig. 1.5. (**a**) Field distribution in the normally conducting and in the ideally conducting state with a homogeneous external magnetic field $\boldsymbol{H}_e$. (**b**) Distribution of the magnetic field in the ideally conducting state after switching off the external magnetic field $\boldsymbol{H}_e$.

A second basic phenomenon of superconductors lies in their magnetic properties. To understand this let us first consider a hypothetical ideal conductor in the magnetic field. Figure 1.5a shows first a non-magnetic real conductor in a homogeneous external field $\boldsymbol{H}_e$. The magnetic flux density $\boldsymbol{B}$ through the body is uniform. If the body is now brought into the state of ideal conductivity, the flux density in the interior at first remains unchanged, because the fluxes enclosed in every possible closed loop in the interior must remain constant, see (1.4). They must then also remain constant if the external magnetic field $\boldsymbol{H}_e$ is switched off, i.e. the distribution of the magnetic flux density in the interior must remain unchanged. This state is shown in Fig. 1.5.b. In the absence of an external magnetic field a homogeneous magnetic flux density can only be maintained in the interior of the body if a surface current flows over the body.

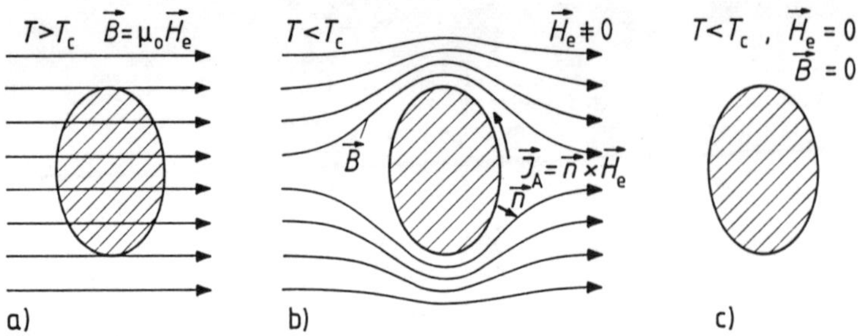

Fig. 1.6a–c. Field distribution, (a) in the normally conducting state with homogeneous external magnetic field $\boldsymbol{H}_e$, (b) in the superconducting state with homogeneous external magnetic field $\boldsymbol{H}_e$, (c) in the superconducting state with external magnetic field switched off

In contrast to the ideal conductor a superconductor shows a different behaviour in this experiment. Magnetic flux is completely excluded from the interior of superconductors. This so-called Meissner effect was discovered in 1933 by *Meissner* and *Ochsenfeld* [1.5]. Figure 1.6 depicts the experiment carried out above, but now with a superconductor. When $T > T_c$ Fig. 1.6.a shows the normally conducting body in a homogeneous external magnetic field which passes uniformly through the body. Now the temperature is reduced below the critical temperature whilst the external magnetic field $\boldsymbol{H}_e$ is maintained. The body becomes superconducting. Because of the Meissner effect the magnetic flux is expelled from the interior. The flux lines in outer space run round the body. Surface currents are caused on the superconductor, in order to satisfy the boundary conditions for the tangential magnetic field strengths outside and inside the superconductor. If now the external magnetic field is switched off, as in Fig. 1.6c, these surface currents also vanish.

The surface currents are in this connection also known as screening currents. The current surface density is according to [1.6] calculated to be

$$\boldsymbol{J}_A = \boldsymbol{n} \times (\boldsymbol{H} - \boldsymbol{H}_\mathrm{i}) \quad , \tag{1.8}$$

where $\boldsymbol{H} = \boldsymbol{B}/\mu_0$ is the magnetic field directly outside the body and $\boldsymbol{H}_\mathrm{i}$ that directly inside the surface. $\boldsymbol{n}$ is the outward normal. With $\boldsymbol{H}_\mathrm{i} = 0$ (1.8) becomes

$$\boldsymbol{J}_A = \boldsymbol{n} \times \boldsymbol{H} \quad . \tag{1.9}$$

This relationship is also valid between the phasors of the surface current density and the magnetic field at the surface of good normal conductors at high frequencies.

1.2 London Equations

We usually describe the connection between the electric field strength $\boldsymbol{E}$, the magnetic field strength $\boldsymbol{H}$ and the electric current density $\boldsymbol{J}$ by the Maxwell equations

$$-\nabla \times \boldsymbol{E} = \dot{\boldsymbol{B}} \quad ; \quad \nabla \times \boldsymbol{H} = \varepsilon \dot{\boldsymbol{E}} + \boldsymbol{J} \quad . \tag{1.10}$$

The current density $\boldsymbol{J}$ at a point is normally determined by the electric field $\boldsymbol{E}$ at this point. In normal conductors we then have

$$\boldsymbol{J} = \sigma \boldsymbol{E} \quad . \tag{1.11}$$

In this section we shall set up a corresponding relationship for superconductors and supplement the Maxwell equations by the London equations. Equation (1.11) is a specialisation of the general relationship

$$\boldsymbol{J} = nQ\boldsymbol{v}_{\mathrm{m}} \quad , \tag{1.12}$$

where n is the density of the mobile portion of the charge carriers, Q their charge and $\boldsymbol{v}_{\mathrm{m}}$ their mean velocity in the direction of $\boldsymbol{J}$.

In what follows we shall denote the density of the superconducting charge carriers by n_{s}, their charge by q_{s}, their velocity by $\boldsymbol{v}_{\mathrm{s}}$ and their mass by m_{s}. Using these quantities the equation of motion of the superconducting charge carriers in collision-free motion becomes

$$m_{\mathrm{s}} \dot{\boldsymbol{v}}_{\mathrm{s}} = q_{\mathrm{s}} \boldsymbol{E} \quad . \tag{1.13}$$

If (1.12) is differentiated with respect to time and then $\dot{\boldsymbol{v}}_{\mathrm{s}}$ is substituted from (1.13) we obtain

$$\dot{\boldsymbol{J}}_{\mathrm{s}} = \frac{n_{\mathrm{s}} q_{\mathrm{s}}^2}{m_{\mathrm{s}}} \boldsymbol{E} \quad . \tag{1.14}$$

From the first of Maxwell's equations it then follows that

$$\nabla \times \dot{\boldsymbol{J}}_{\mathrm{s}} = -\frac{1}{\mu_0 \lambda_{\mathrm{L}}^2} \dot{\boldsymbol{B}} \tag{1.15}$$

with the London penetration depth

$$\lambda_{\mathrm{L}} = \sqrt{\frac{m_{\mathrm{s}}}{\mu_0 n_{\mathrm{s}} q_{\mathrm{s}}^2}} \quad . \tag{1.16}$$

From the second of Maxwell's equations, with $\boldsymbol{B} = \mu_0 \boldsymbol{H}$ and (1.15) we have, neglecting the displacement current density $\varepsilon \dot{\boldsymbol{E}}$,

$$\nabla \times (\nabla \times \dot{\boldsymbol{B}}) = -\frac{1}{\lambda_{\mathrm{L}}^2} \dot{\boldsymbol{B}} \quad . \tag{1.17}$$

The double curl of a vector can now be replaced according to the calculation rules of vector analysis by the Laplace operator Δ

$$\nabla \times (\nabla \times \boldsymbol{v}) = \nabla(\nabla \boldsymbol{v}) - \Delta \boldsymbol{v} \quad . \tag{1.18}$$

If one takes into account that the magnetic field is source-free, and hence $\nabla B = 0$, we find from (1.17) and (1.18) that

$$\Delta \dot{\boldsymbol{B}} = \frac{1}{\lambda_{\mathrm{L}}^2} \dot{\boldsymbol{B}} \quad . \tag{1.19}$$

The significance of this equation can best be made clear by a one-dimensional case. We therefore assume that $\dot{\boldsymbol{B}}$ depends only on the x-coordinate. Then (1.19) becomes

$$\frac{\partial^2 \dot{\boldsymbol{B}}}{\partial x^2} = \frac{1}{\lambda_{\mathrm{L}}^2} \dot{\boldsymbol{B}} \quad . \tag{1.20}$$

Of the two particular solutions, which are proportional to $\exp(-x/\lambda_{\mathrm{L}})$ and to $\exp(x/\lambda_{\mathrm{L}})$, for the problem illustrated in Fig. 1.7 only the solution with the negative exponent is physically meaningful

$$\dot{\boldsymbol{B}}(x) = \dot{\boldsymbol{B}}_{\mathrm{e}} \mathrm{e}^{-x/\lambda_{\mathrm{L}}} \quad . \tag{1.21}$$

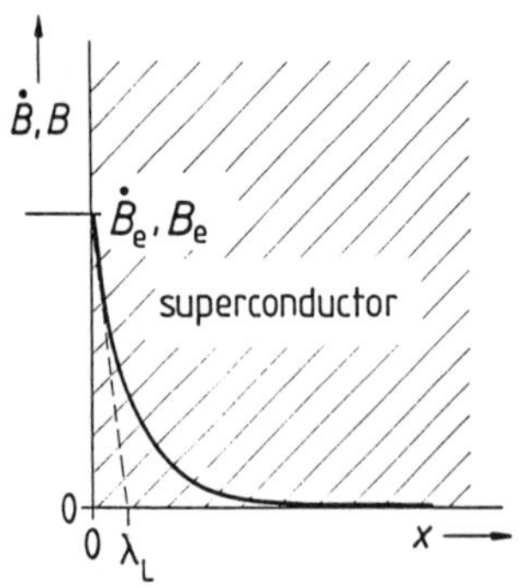

Fig. 1.7. Penetration of the time derivative of the magnetic flux density $\dot{\boldsymbol{B}}$ and of the magnetic flux density $\boldsymbol{B}$ itself in a superconductor

The timewise derivative of the magnetic flux density in the interior of a superconductor therefore falls off exponentially inwards. The decay constant setting the scale is the London penetration depth λ_{L}

From the Meissner effect we know that deep in the interior of a superconductor not only does the time derivative of the magnetic flux density vanish, but so also does the magnetic flux density itself. *F.* and *H. London* [1.7] therefore proposed to apply (1.19) not only to $\dot{\boldsymbol{B}}$, but also to $\boldsymbol{B}$

$$\Delta \boldsymbol{B} = \frac{1}{\lambda_{\mathrm{L}}^2} \boldsymbol{B} \quad . \tag{1.22}$$

Then for example in the one-dimensional problem of Fig. 1.7 the magnetic flux density also shows the position dependence

$$B(x) = B_{\hat{e}} e^{-x/\lambda_L} \quad . \tag{1.23}$$

If we trace back the argument by which we derived (1.19) it turns out that we obtain (1.22) if the relationship (1.15) also holds between the quantities J_s and B, not differentiated with respect to time,

$$\nabla \times J_s = -\frac{1}{\mu_0 \lambda_L^2} B \quad . \tag{1.24}$$

This is the first London equation. The second follows directly from (1.14):

$$\dot{J}_s = \frac{1}{\mu_0 \lambda_L^2} E \quad . \tag{1.25}$$

Here μ has been set equal to μ_0 for the interior of the superconductor. According to the two-fluid model the current density J in the superconductor can be regarded as the superposition of the superconducting current J_s on a normally conducting current J_n

$$J = J_n + J_s \quad . \tag{1.26}$$

Here we have

$$J_n = \sigma_1 E \quad . \tag{1.27}$$

In the steady state with sinusoidal time dependence the description with phasors is useful. These are time-invariant complex quantities. From them the time-dependent functions are obtained by multiplying by $\exp(j\omega t)$ (ω = circular frequency) and taking the real part. Accordingly $E(t) = \sqrt{2}\,\mathrm{Re}\{\underline{E}\exp(j\omega t)\}$: usually a factor $\sqrt{2}$ is introduced, so that $\underline{E}$ represents an r.m.s. value. If vectors are written as phasors, then it should be understood that their components are phasors. So in the alternating state it follows from (1.25) to (1.27) for the phasors of the current density and the electrical field strength at frequency ω that

$$\underline{J} = (\sigma_1 - j\sigma_2)\underline{E} \tag{1.28}$$

with

$$\sigma_2 = \frac{1}{\omega \mu_0 \lambda_L^2} \quad . \tag{1.29}$$

The experimentally determined penetration depth λ is as a rule greater than the London penetration depth λ_L determined from (1.16), see also Sect. 1.3.

On account of the temperature dependence of the density n_s of the superconducting electrons the penetration depth is also temperature dependent. Experimental results are well represented by

$$\frac{\lambda(T)}{\lambda(0)} = \frac{1}{\sqrt{1-(T/T_c)^4}} \quad . \tag{1.30}$$

The penetration depth therefore becomes greater with increasing temperature. As $T \to T_c$ it becomes infinitely large. The penetration depth $\lambda(0)$ of typical superconductors lies in the region of 10 to 1000 nm (see Table 1.1).

The London theory is a simple phenomenological theory. It allows many superconductor phenomena to be qualitatively described correctly. The London equations, however, do not have the same status as, for example, Maxwell's equations, which one accepts as exact.

The London theory does not always provide results with adequate quantitative accuracy. An example of this is the calculation of the current heat losses in superconductors, see Sect. 1.4. The London theory is always less reliable when strong magnetic fields or very thin superconducting layers are present.

1.3 Cooper Pairs and the Energy Band Model

The BCS theory of *Bardeen, Cooper* and *Schriefer* [1.8] provides a more detailed description of observational results on superconductors than the London equations. The most important results of this microscopic theory of superconductivity, together with an energy band model for the electrons in superconductors, will be described in this section.

Between the free electrons in a metal there normally exist Coulomb repulsion forces, which ensure that the electrons do not approach each other too closely. The BCS theory now assumes that in a superconductor additional attractive forces exist, which compensate for the repulsive forces. As a result, pair formation between electrons becomes energetically advantageous, and hence probable.

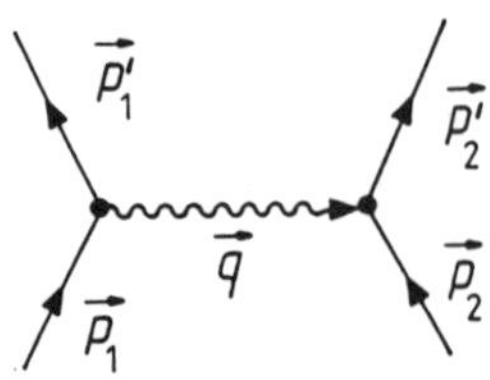

Fig. 1.8. Interaction of two electrons in the exchange of a phonon. $\boldsymbol{p}_1, \boldsymbol{p}_2$ are the momenta of the electrons before, and $\boldsymbol{p}_1', \boldsymbol{p}_2'$ are the momenta of the electrons after the exchange of the phonon with the momentum $\boldsymbol{q}$

Such attractive forces can arise when two electrons exchange a phonon, which is a quantum of thermal energy in the lattice vibrations. This exchange is illustrated schematically in Fig. 1.8. It is a process in which the sum of the momenta remains the same.

The average distance at which the attractive and repulsive forces compensate is called the coherence length ξ_{co}. It is accordingly the average separation distance between two electrons forming a so-called Cooper pair. ξ_{co} lies in the order of magnitude of 0.1 to 1 μm. The two electrons of a Cooper pair have opposite momenta $\boldsymbol{p}_1 = -\boldsymbol{p}_2$ and opposite spin. The pair can therefore be characterised by $(\boldsymbol{p}\uparrow, -\boldsymbol{p}\downarrow)$. With the wave number vector $\boldsymbol{k}$ and the relationship between the particle model and the wave model

$$\boldsymbol{p} = \hbar \boldsymbol{k} \tag{1.31}$$

the pair can also be characterised by

$$(\boldsymbol{k}\uparrow, -\boldsymbol{k}\downarrow) \quad . \tag{1.32}$$

These representations are valid first of all only for vanishing superconducting current density, since with $\boldsymbol{p}_1 = -\boldsymbol{p}_2$ the momentum of the Cooper pair is equal to zero and there is no transport of charge.

The extent of a Cooper pair is much greater than the average separation of two conduction electrons. In the region of a pair there are 10^6 to 10^7 other electrons, which are themselves correlated in pairs. According to the BCS theory this correlation now as a result exists not only between the two electrons of a pair, but also between all the Cooper pairs of a superconductor. In the quantum mechanical sense they all have the same state. They do not need to obey the Pauli exclusion of multiple occupation of the same state because their total momentum is zero. The description of the totality of all the Cooper pairs of a superconductor is therefore possible by only one wave function:

$$\psi(\boldsymbol{r}) = |\psi(\boldsymbol{r})| \, \mathrm{e}^{\mathrm{j}\theta(\boldsymbol{r})} \quad . \tag{1.33}$$

Here $\boldsymbol{r}$ is the position vector and θ the phase. $|\psi(\boldsymbol{r})|^2$ is proportional to the density n_c of the Cooper pairs.

The phase coherence between the Cooper pairs described by (1.33) extends throughout the whole superconductor. It can extend, for example in the case of a superconducting coil, for several kilometres. This phase coherence leads to macroscopically observable quantum effects, which include the flux quantisation described in Sect. 1.5.

In the superconductor there are, in addition to the superconducting electrons bound in Cooper pairs, also normally conducting mobile electrons. In what follows an energy band model will be developed for both, similar to that which is familiar for electrons and holes in the semiconductor.

We consider first of all in Fig. 1.9a the dependence of the energy on the wave number for an electron in the normal conductor. In the neighbourhood of the edge of the conduction band W_L the behaviour is roughly quadratic [1.8.]. The Fermi level $W_F > W_L$ lies in the conduction band. In order to take an electron from state β into state α one needs the energy $W_\alpha - W_\beta$. This energy can be partitioned in the following way:

$$W_\alpha - W_\beta = (W_\alpha - W_F) + (W_F - W_\beta) \quad . \tag{1.34}$$

The portion $W_F - W_\beta$ can be regarded as the energy needed to remove the electron at W_β thus producing a hole. The portion $W_\alpha - W_F$ is then the energy required to occupy the state α with an electron. Using the abbreviations

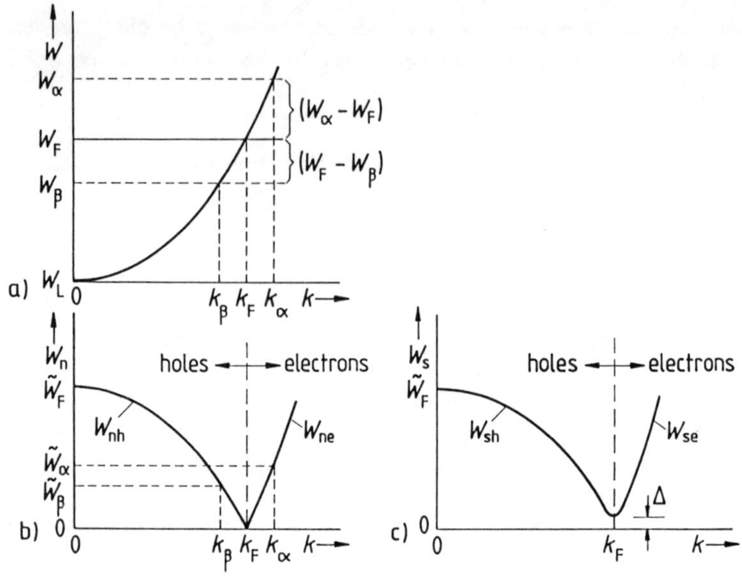

Fig. 1.9a–c. Energies as a function of wave number k; (k_{F}) wave number at the Fermi energy W_{F}. **(a)** electron energy W in the normal conductor, (W_{L}) edge of conduction band. **(b)** Reduced electron energy $\tilde{W}_{\mathrm{e}}$ and hole energy $\tilde{W}_{\hat{\mathrm{h}}}$ in the normal conductor. **(c)** $\tilde{W}_{\mathrm{e}}$ and $\tilde{W}_{\mathrm{n}}$ for individual electrons in the superconductor; (Δ) energy gap

$$\tilde{W}_\alpha = W_\alpha - W_{\mathrm{F}}\,, \quad \tilde{W}_\beta = W_{\mathrm{F}} - W_\beta\,, \quad \tilde{W}_{\mathrm{F}} = W_{\mathrm{F}} - W_{\mathrm{L}} \tag{1.35}$$

we can represent the reduced electron- and hole-energies of the normal conductor

$$W_{\mathrm{ne}} = W - W_{\mathrm{F}}\,, \quad W_{\mathrm{nh}} = W_{\mathrm{F}} - W \tag{1.36}$$

as shown in Fig. 1.9.b. In the ground state all possible values of the wave number k below k_{F} are occupied and above k_{F} are unoccupied. Then arbitrarily small energies suffice for an excitation, i.e. for the creation of a hole below k_{F} and for the occupation of a state above k_{F} by an electron.

The corresponding diagram for the individual electrons of a superconductor is shown in Fig. 1.9c. At $k = k_{\mathrm{F}}$, W_{s} does not go to zero, but there remains an energy gap Δ. This means that an excitation of an individual electron requires an energy of at least 2Δ. Roughly speaking, Δ lies in the order of magnitude of 1 to 10 meV, see also Table 1.1. Accordingly for superconductors, as the frequency increases, a photon absorption first occurs in the millimetre wave region. It should be pointed out that Fig. 1.9c is not drawn to scale. $\tilde{W}_{\mathrm{F}}$ is several orders of magnitude larger than Δ.

Just as Fig. 1.9 shows the energies of the states, Fig. 1.10 shows their occupation probabilities. They are shown first of all for the ground state, i.e. for $T = 0\,\mathrm{K}$. According to Fig. 1.10a the Fermi distribution in the normal conductor is a rectangular function. One excitation from the ground state is illustrated. Figure 1.10b shows the occupation probability for a superconductor being in the ground state except for four excitations. In the ground state all the electrons are

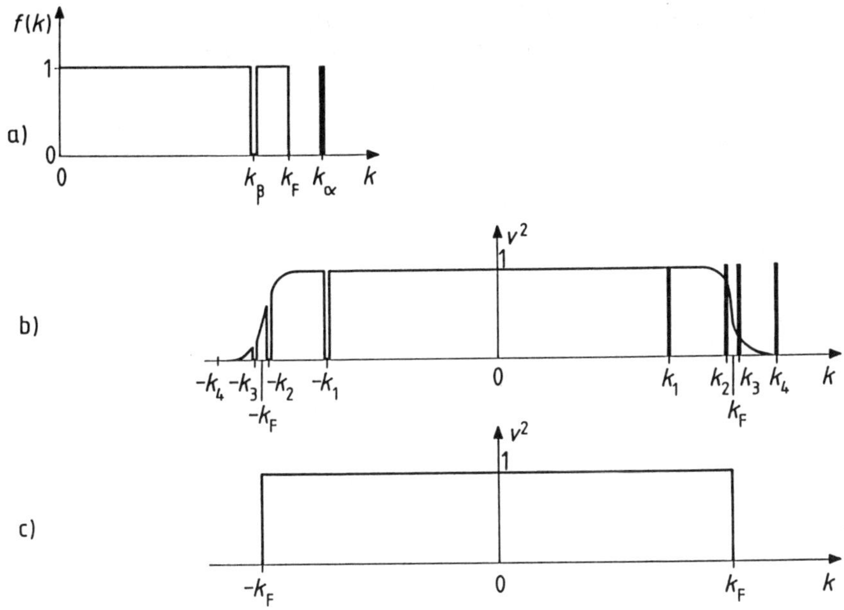

Fig. 1.10a–c. Occupation probabilities as functions of the wave number k, temperature $T = 0\,\text{K}$. **(a)** Normal conductor with one excitation. **(b)** Probability v^2 of the state occupation by Cooper pairs, 4 excitations. **(c)** Probability $v^2(k)$ approximated by a rectangular function

bound to Cooper pairs ($\boldsymbol{k}\uparrow, -\boldsymbol{k}\downarrow$). The occupation probability $v^2(k)$ is therefore symmetric with respect to $k = 0$. It is, however, not a rectangular function but has a continuous variation in the region of k_F. Since excitations in the superconductor take both positive and negative k-values, the occupation probability is shown also for negative k-values. An excitation of the state k is defined as

$$\boldsymbol{k}\uparrow \text{ certainly occupied; } -\boldsymbol{k}\downarrow \text{ certainly empty} \quad . \tag{1.37}$$

By this excitation the momentum of the system is raised by $\boldsymbol{p} = \hbar\boldsymbol{k}$.

Figure 1.10b illustrates not only the ground state but also four excitations from the ground state. They will now be described in somewhat more detail.

Excitation 1: $v^2(k_1)$ was previously approximately 1. The excitation causes no change. $v^2(-k_1)$ was previously approximately 1, and after the excitation is zero. In total there is a decrease in probability of about 1. The excitation acts as a hole.

Excitation 2: Before the excitation $v^2(k_2)$ was about 0.7. With the excitation there is an increase of 0.3. $v^2(-k_2)$ before the excitation was 0.7, after the excitation there is a decrease of 0.7. In total we have a decrease in the occupation probability of about 0.4. The excitation is hole-like.

In corresponding manner the excitations k_3 and k_4 are shown to be electron-like and an electron, respectively. Because their character is not always completely hole-like or electron-like, the excitations are called quasi-particles: either quasi-electrons or quasi-holes according to their predominant nature.

The actual behaviour of $v^2(k)$ in the ground state around $k = \pm k_F$ is much steeper than appears in Fig. 1.10b. It is therefore often sufficient to approximate the occupation probability for Cooper pairs in the ground state by a rectangular function as in Fig. 1.10c.

When $T > 0$ the occupation probability in the normal conductor is described by the Fermi distribution. This Fermi distribution can now also be used as a basis in calculating the thermal excitation ($T > 0$) of individual electrons in the superconductor from the ground state.

For the calculation of the occupation density of the states in the superconductor one must also know the density of the states. Since this is determined in the first place by the nature of the lattice structure, one starts off with the assumption that in the transition from the normally conducting to the superconducting state the density of the states as a function of the wave number k is unaltered. Accordingly $(\mathrm{d}N/\mathrm{d}k)\mathrm{d}k$ remains constant, i.e.

$$\frac{\mathrm{d}N}{\mathrm{d}W_s}\frac{\mathrm{d}W_s}{\mathrm{d}k}\mathrm{d}k = \frac{\mathrm{d}N}{\mathrm{d}W_n}\frac{\mathrm{d}W_n}{\mathrm{d}k}\mathrm{d}k \quad . \tag{1.38}$$

Denoting the density of the states in the normal conductor by $D_n(W_n) = \mathrm{d}N/\mathrm{d}W_n$, and the density of the states in the superconductor by $D_s(W_s) = \mathrm{d}N/\mathrm{d}W_s$, we find from (1.38) that

$$D_s = D_n \frac{\mathrm{d}W_n}{\mathrm{d}k}\frac{\mathrm{d}k}{\mathrm{d}W_s} \quad . \tag{1.39}$$

Comparison of Figs. 1.9b and 1.9c shows that when $k \simeq k_F$ the following assumption seems to hold:

$$W_s(k) = \sqrt{\Delta^2 + W_n^2(k)} \quad . \tag{1.40}$$

Hence, from (1.39),

$$D_s(W_s) \simeq D_n(W_n = 0)\frac{W_s}{\sqrt{W_s^2 - \Delta^2}} \quad \text{for} \quad |W_s| \geq \Delta \quad . \tag{1.41}$$

The behaviour of D_s is interesting especially for small values of W_s, and hence for $k \simeq k_F$. For these values D_n depends only weakly on k and on W_n. Accordingly, the value of D_n at the point $W_n = 0$ has already been substituted as a simplification in (1.41). D_s tends to infinity as W_s tends to Δ. When $|W_s| < \Delta$ then $D_s = 0$:

$$D_s(W_s) = 0 \quad \text{for} \quad |W_s| < \Delta \quad . \tag{1.42}$$

We thus obtain for the superconductor in the ground state the dependence of the density of states D_s on the energy W_s as in Fig. 1.11a. For temperatures $T > 0$ the hole states and the electron states are distributed according to the

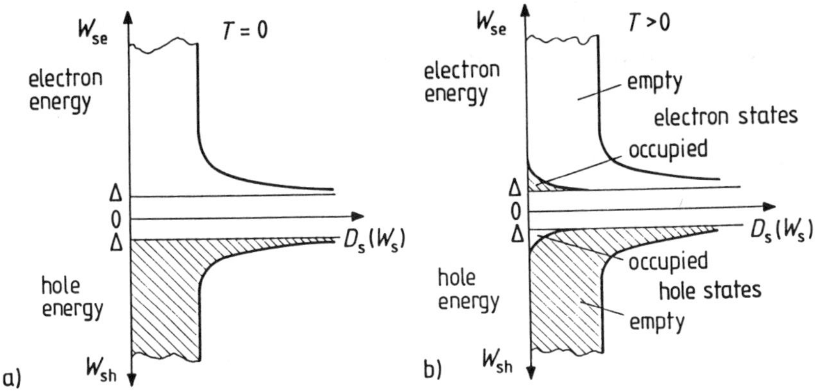

Fig. 1.11a,b. Density of states D_s as a function of the electron and hole energies W_{se} and W_{sh} in the superconductor. Shaded regions are states occupied by electrons. **(a)** $T = 0$. **(b)** $T > 0$

Fermi function. In Fig. 1.11b the shaded areas are an indication of the states then occupied by electrons.

So far we have been considering the current-free state of the superconductor. An imposed direct current with the density $J > 0$ is first of all carried entirely by the Cooper pairs with their total momentum $\boldsymbol{p}_\infty$. In this situation all the Cooper pairs remain in the same quantum state.

When a current is carried by a familiar type of conductor, e.g. by an ordinary conductor or a semiconductor, there inevitably occurs a resistance, because the mobile charge carriers are scattered. Their momenta are therefore always changing, so that their free acceleration in the direction of the electric field is hindered. The scattering can be caused by thermal lattice vibrations or by disturbances in the otherwise periodic crystal structure. In the case of the superconductor the two electrons forming a Cooper pair are indeed constantly pushing one another as in Fig. 1.8, but their total momentum remains constant, and the current flux does not vary. Only a scattering process which changes the total momentum in the direction of the current flux can reduce the current flux. That can only happen if the pair is broken. This breakage, however, requires at least the energy 2Δ, see Fig. 1.9c. A current reducing scattering process can only occur if this energy is supplied from somewhere. Small current densities cannot transfer this energy to the Cooper pair. Accordingly in these circumstances there are no scattering processes which change the total momentum. There is then no electrical resistance.

If the current density increases it eventually becomes energetically possible for the Cooper pairs to be broken into individual electrons. The superconductor goes over into the normally conducting state and acquires a resistance. Associated with this critical current density are maximum currents which can be carried along wires in the superconducting state. For example, the critical current along a lead wire of 1 mm diameter at a temperature of 4.2 K is about 100 A.

As we have already seen in Sect. 1.1, the current density in the outer skin of a superconductor is closely linked to the magnetic field strength outside the superconductor. A superconductor therefore also goes over into the normally conducting state if an external applied field exceeds the so-called critical magnetic field strength. For lead at 4.2 K this value is about 4×10^4 A/m.

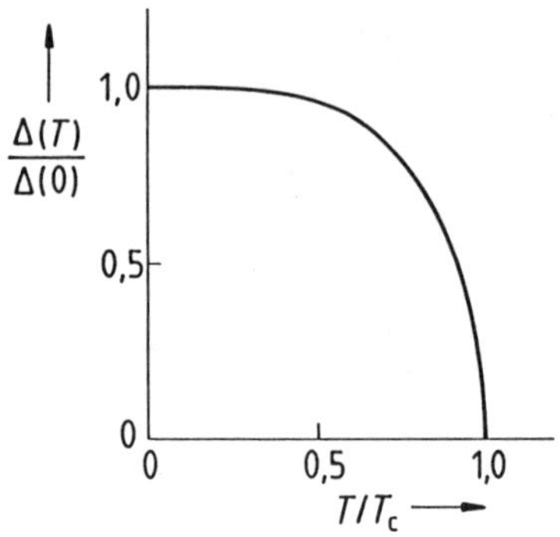

Fig. 1.12. Dependence of the energy gap Δ on the temperature T according to the BCS theory, (T_c) critical temperature

According to the BCS theory the disappearance of superconduction above the critical temperature T_c is because the energy gap Δ vanishes when $T > T_c$. Figure 1.12 shows how the energy gap depends on the temperature according to the BCS theory. This behaviour can be represented to a good approximation by

$$\frac{\Delta(T)}{\Delta(0)} \simeq \sqrt{\cos\left[\frac{\pi}{2}\left(\frac{T}{T_c}\right)^2\right]} \quad . \tag{1.43}$$

According to the BCS theory, for every superconductor the energy gap at 0 K is proportional to the critical temperature:

$$2\Delta(0\,\mathrm{K}) = 3.52 k_B T_c \quad . \tag{1.44}$$

Deviations from (1.44) observed experimentally are usually within $\pm 30\%$.

With the help of the coherence length ξ_{co} one can now make corrections to the London penetration depth λ_L given by (1.16), in order to achieve agreement with experimental findings. There are three cases to distinguish [1.3].

1) In a few pure metals $\xi_{co} \ll \lambda_L$. In this case λ_L does not need modification.

$$\lambda = \lambda_L \quad \text{for} \quad \xi_{co} \ll \lambda_L \quad . \tag{1.45}$$

2) For pure superconductors with $\xi_{co} \gg \lambda_L$ the penetration depth becomes

$$\lambda = 0.65 \lambda_L \left(\frac{\xi_{co}}{\lambda_L}\right)^{1/3} \quad \text{for} \quad \xi_{co} \gg \lambda_L \quad . \tag{1.46}$$

3) In materials with impurities and in alloys the free path length l of the electrons is shorter than in the pure material. The coherence length is also reduced. For small free path lengths the penetration depth is then given by

$$\lambda = \lambda_L \sqrt{\frac{\xi_{co0}}{\xi_{co}}} \simeq \lambda_L \sqrt{\frac{\xi_{co0}}{l}} \quad \text{for} \quad l^3 \ll \xi_{co} \lambda_L^2 \quad . \tag{1.47}$$

The connection with the coherence length ξ_{co0} of the pure material is found experimentally to be

$$\frac{1}{\xi_{co}} = \frac{1}{\xi_{co0}} + \frac{1}{\alpha l}; \quad \alpha \simeq 1 \quad . \tag{1.48}$$

For example, the penetration depth for tin with indium impurities depends on the free path length as shown in Fig. 1.13.

The temperature dependence of the penetration depth is still given to a good approximation by (1.30).

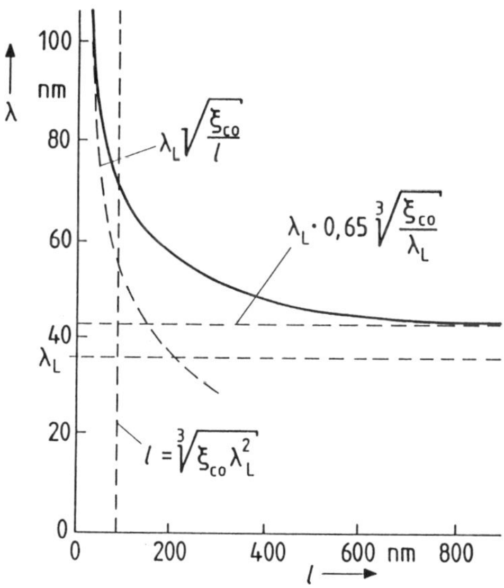

Fig. 1.13. Penetration depth λ as a function of the free path length l of the electrons, determined by indium impurities in tin. $T = 0\,\mathrm{K}$, $\xi_{co0} = 6.4\lambda_L$

1.4 Current Heat Losses in Normal Conductors and Superconductors

Current heat losses in normal conductors are caused by the skin effect. According to Fig. 1.14 current density and field strength fall off exponentially in the interior of the conductor. Here the classical penetration depth δ_c is the real part of the reciprocal of the propagation constant in the wall with the permeability μ_0 and the permittivity $-j\sigma/\omega$. Here σ is the specific conductivity of the wall material [1.10]

$$\delta_c = \sqrt{2/(\omega\mu_0\sigma)} \quad . \tag{1.49}$$

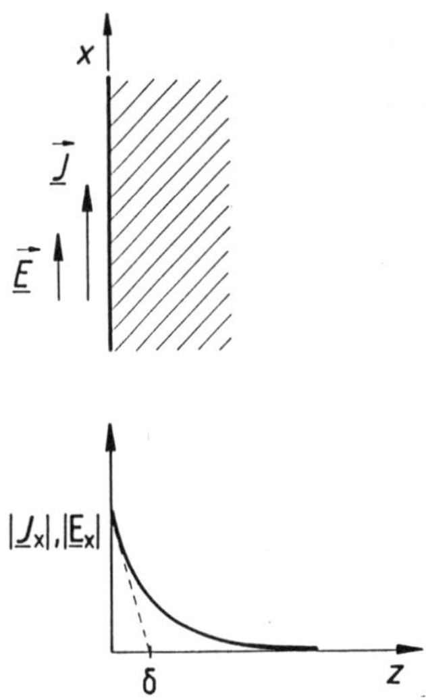

Fig. 1.14. Electric field strength $\underline{E}_x$ and volume current density $\underline{J}_x$ at the surface of a conductor

The surface impedance

$$Z_s = \underline{E}_x(z=0) \Big/ \int_0^\infty \underline{J}_x(z)\mathrm{d}z \tag{1.50}$$

can be calculated in similar manner from the wave impedance of the wall material:

$$Z_s = \sqrt{\mathrm{j}\omega\mu_0/\sigma} \quad . \tag{1.51}$$

Its real part

$$R_s = \sqrt{\omega\mu_0/(2\sigma)} \tag{1.52}$$

describes the current heat loss. Equation (1.51) was derived with $\boldsymbol{J} = \sigma\boldsymbol{E}$. It was therefore assumed that the current density at a point depends only on the strength of the electric field at that point $\boldsymbol{r}$. This is correct, so long as the classical skin penetration depth δ_c from (1.49) is very much greater than the mean free path length l of the electrons. In general l increases with cooling, so that for many normal conductors in the microwave region at low temperatures this condition for the classical skin effect is no longer fulfilled. The volume current density $\boldsymbol{J}$ then depends on the distribution of the electric field strength in a region round about $\boldsymbol{r}$, and in fact satisfies the relation [1.18]

$$\boldsymbol{J}(\boldsymbol{r}) = \frac{3\sigma}{4\pi l} \int \boldsymbol{\varrho}[\boldsymbol{\varrho} \cdot \boldsymbol{E}(\boldsymbol{r} + \boldsymbol{\varrho})]\varrho^{-4} \mathrm{e}^{-\varrho/l} dV' \quad . \tag{1.53}$$

Here dV' is an infinitesimally small volume element at the location $\boldsymbol{r} + \boldsymbol{\varrho}$. The equation (1.53) is now to be substituted in the Maxwell equations. In the case of the pronounced anomalous skin effect ($l \gg \delta_c$) one then obtains for the surface resistance

$$R_s = \left(\frac{3\omega\mu_0}{4}\right)^{2/3} \left(\frac{l}{\sigma}\right)^{1/3} \quad . \tag{1.54}$$

The anomalous skin penetration depth is

$$\delta_a = \left(\frac{2\sqrt{3}}{\pi}\delta_c^2 l\right)^{1/3} \tag{1.55}$$

and hence greater than the classical penetration depth δ_c. The surface resistance R_s from (1.54) depends only on the ratio l/σ, which is a nearly temperature-independent material constant. For $l \gg \delta_c$, R_s is therefore independent of the temperature and also independent of a more or less lucky estimate for the conductivity. Figure 1.15 shows the calculated surface resistance as a function of the frequency for copper at room temperature and at 4.2 K.

At low frequencies the curves start off horizontally, because in these diagrams it was assumed that the conductor is only about 1 μm thick. Experiments [1.11] confirm that at 10 GHz the surface resistance cannot even be reduced to one tenth whilst cooling from room temperature to 4.2 K. These measurements were carried out with microstrip resonators made from rolled copper deposits on glass-fibre reinforced Teflon substrates. Resonators with electrolytically deposited copper showed a still smaller improvement.

To calculate the current heat losses in the superconductor one can start from the complex conductivity according to (1.28):

$$\sigma = \sigma_1 - \mathrm{j}\sigma_2 \quad . \tag{1.56}$$

With well developed superconduction $\sigma_2 \gg \sigma_1$, and from (1.51), R_s is given by

$$R_s \approx \frac{0.5\sigma_1\sqrt{\omega\mu_0}}{\sigma_2^{3/2}} \quad . \tag{1.57}$$

σ_2 follows from (1.29) with λ_L replaced by λ. To determine σ_1 one assumes that σ_1 is proportional to the ratio of the normally conducting electrons to the total

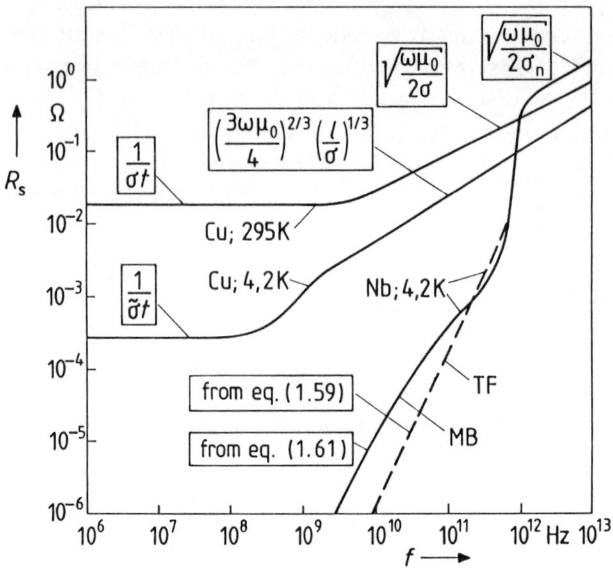

Fig. 1.15. Real part of the surface impedance for a film of copper $t = 1\mu$m thick at 295 K and 4.2 K, and also of niobium at 4.2 K. at 295 K: $\sigma_{Cu} = 0.059 \times 10^4 \times 1/\Omega$cm, $l_{Cu} = 38$nm; at 4.2 K:$\sigma_{cu} = 29.4 \times 10^4 \times 1/\Omega$cm, $l_{Cu} = 19.1$nm; $\tilde{\sigma}$ – effective conductivity at $t < l_{Cu}$.$\sigma_{nNb} = 0.0157 \times 10^4 \times 1/\Omega$cm, $l_{Nb} = 11$nm, $\lambda_{Nb} = 86$nm; (TF) two fluid model, (MB) Mattis-Bardeen theory. From [1.13]

number of the electrons contributing to the conduction. From (1.30) and (1.16) it follows that this ratio is proportional to $(T/T_c)^4$ when $T < T_c$. Accordingly we write

$$\sigma_1 = \sigma_n \left(\frac{T}{T_c}\right)^4 \tag{1.58}$$

where σ_n is the conductivity at T_c.

From (1.57), with (1.58), (1.30) and (1.29), it follows that

$$R_s = \frac{0.5\mu_0^2\sigma_n\lambda^3(0)\omega^2(T/T_c)^4}{[1 - (T/T_c)^4]^{3/2}} \ . \tag{1.59}$$

To this approximation the surface resistance is accordingly proportional to the square of the frequency. The broken line in Fig. 1.15 shows the surface resistance of niobium at 4.2 K according to (1.59).

Just as in normal conductors the local relationship requires the fulfilment of the condition that $l \ll \delta_c$, so in the two-fluid model calculation described above for superconductors we must have the condition $\xi_{co} \ll \lambda$. This ensures that the field does not change over the extent of a Cooper pair, see also (1.45). This condition, however, is fulfilled for only a few superconductors. In general a non-local relationship is required, as in (1.53). In the Mattis-Bardeen theory [1.12] the corresponding calculations are carried out on the basis of the BCS theory. The results can be presented simply only in the limiting cases. For $\lambda \ll \xi_{co}$,

$\omega \ll 2\Delta/\hbar$ and $k_B T \ll \Delta$ explicit expressions can be given for the real part and the imaginary part of the complex conductivity. For σ_2 (1.29) is again valid, and σ_1 is

$$\sigma_1 = \sigma_n \frac{2\Delta}{(k_B T)} e^{-\Delta/(k_B T)} \ln \frac{\Delta}{(\hbar\omega)} \quad . \tag{1.60}$$

This intermediate result of a non-local calculation can now be substituted into the equation (1.57) derived under local conditions. In spite of this inconsistency the surface resistance calculated in this way often describes the experimental results quite well:

$$R_s = \frac{0.5\mu_0^2 \sigma_n \lambda^3(0)}{[1-(T/T_c)^4]^{3/2}} \frac{2\Delta}{(k_B T)} e^{-\Delta/(k_B T)} \omega^2 \ln \frac{\Delta}{(\hbar\omega)} \ . \tag{1.61}$$

The fourth curve in Fig. 1.15 shows the surface resistance of niobium at 4.2 K calculated in this way. Niobium has small conductivity, and even at a temperature of 4.2 K its surface resistance above the band gap frequency $f_g = 2\Delta/h = 720$ GHz is already greater than that of copper at room temperature. The two-fluid model leads at lower frequencies to surface resistances which, judged by these results of the Mattis-Bardeen theory, are about an order of magnitude too low.

The calculation of the surface resistance from the general Mattis-Bardeen theory is somewhat complicated, and therefore practical only in individual cases. A calculation of this sort was carried out for a transmission line resonator with the stratification PbAu/SiO/PbInAu. The results for the attenuation constant α caused by current heat losses are plotted as continuous curves for two frequencies in Fig 1.16. Dielectric losses can be neglected because of the small dielectric loss factor of SiO at low temperatures. On account of the extremely small dielectric thickness h the radiation losses are also negligibly small. The calculated curves [1.14] agree very closely with the measured values indicated in Fig. 1.16 by crosses. The measured results were obtained by Q-factor measurements at transmission line resonators. Here the transmission line with the cross-section stated in Fig. 1.16 was brought into a rectangular waveguide thus forming an antipodal fin-line. The broken curves in Fig. 1.16 represent the attenuation constant calculated according to (1.61). So far as can be seen in Fig. 1.16, they deviate more from the correct values, the higher the operating frequency and the lower the temperature.

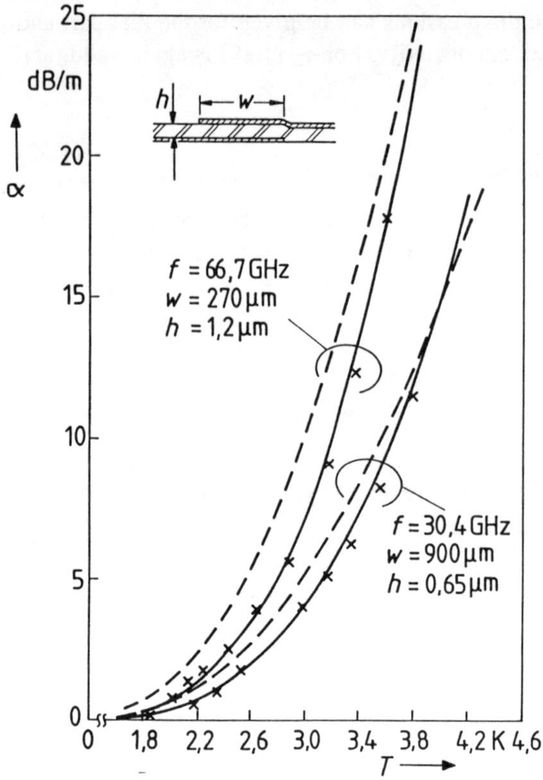

Fig. 1.16. Attenuation constant α of superconducting thin film transmission lines of Pb-Au/SiO/Pb-In-Au as a function of the temperature T; (- - -) from (1.61), (–) by exact evaluation of the Mattis-Bardeen theory, (×) experimental values

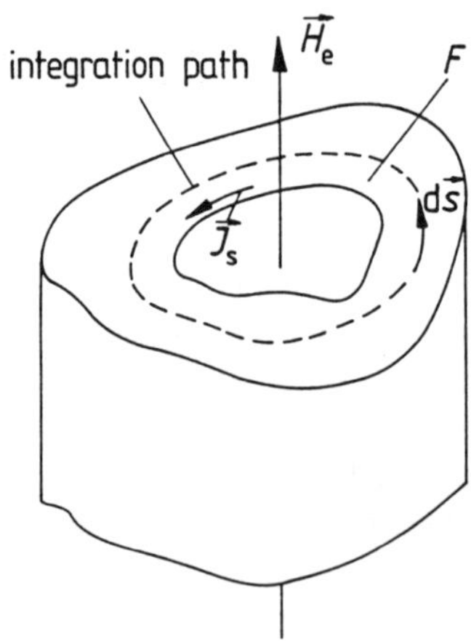

Fig. 1.17. Closed superconducting loop illustrating the flux quantisation

1.5 Flux Quantisation

We have already indicated in Sect. 1.1 that the magnetic flux through every closed superconducting loop must be quantised. We can now understand this if we recall the wave model of the Cooper pairs. Because of the phase coherence according to (1.33), after a closed loop like that in Fig. 1.17, the wave function ψ must return to the same value. Accordingly θ and hence the integral over the wave number can only have changed by an integral multiple of 2π. It then follows from (1.31) for the momentum $\boldsymbol{p}_{\mathrm{co}}$ of the Cooper pairs

$$\oint \boldsymbol{p}_{\mathrm{co}} \mathrm{d}\boldsymbol{s} = nh\,, \quad n = \ldots, -2, -1, 0, 1, 2, \ldots \quad . \tag{1.62}$$

This condition is identical with the quantum condition of the Bohr model of the atom. In (1.62) we have to substitute the so-called canonical momentum $\boldsymbol{p}_{\mathrm{co}}$ of the Cooper pairs. In a magnetic field this canonical momentum of a particle with the charge q_s consists not only of the kinetic momentum $m\boldsymbol{v}$, but also of the contribution $\mu_0 q_\mathrm{s} \boldsymbol{A}$, see e.g. [1.15]. Here the vector potential $\boldsymbol{A}$ is connected with the magnetic field $\boldsymbol{H}$ by the relation

$$\nabla \times \boldsymbol{A} = \boldsymbol{H} \quad . \tag{1.63}$$

According to these rules of quantum mechanics we have

$$\boldsymbol{p}_{\mathrm{co}} = m_\mathrm{s} \boldsymbol{v}_\mathrm{s} + \mu_0 q_\mathrm{s} \boldsymbol{A} \quad . \tag{1.64}$$

Here m_s is the mass, $\boldsymbol{v}_\mathrm{s}$ the velocity and q_s the charge of the Cooper pairs. From (1.62) it follows with (1.64) that

$$nh = \oint m_\mathrm{s} \boldsymbol{v}_\mathrm{s} \cdot \mathrm{d}\boldsymbol{s} + \mu_0 q_\mathrm{s} \oint \boldsymbol{A} \cdot \mathrm{d}\boldsymbol{s} \tag{1.65}$$

Using (1.12) we can introduce the current density $\boldsymbol{J}_\mathrm{s}$ and obtain

$$nh = q_\mathrm{s} \left[\oint \frac{m_\mathrm{s}}{n_\mathrm{s} q_\mathrm{s}^2} \boldsymbol{J}_\mathrm{s} \cdot \mathrm{d}\boldsymbol{s} + \mu_0 \oint \boldsymbol{A} \cdot \mathrm{d}\boldsymbol{s} \right] \quad . \tag{1.66}$$

Now, by virtue of (1.63) and Stokes's law for the magnetic flux φ_F through the integration loop of the surface $\boldsymbol{F}$,

$$\mu_0 \oint \boldsymbol{A} \cdot \mathrm{d}\boldsymbol{s} = \mu_0 \iint_\mathrm{F} \boldsymbol{H} \cdot \mathrm{d}\boldsymbol{F} = \Phi_\mathrm{F} \quad . \tag{1.67}$$

The quantisation condition accordingly takes the form

$$\frac{nh}{q_\mathrm{s}} = \oint \frac{m_\mathrm{s}}{n_\mathrm{s} q_\mathrm{s}^2} \boldsymbol{J}_\mathrm{s} \cdot \mathrm{d}\boldsymbol{s} + \Phi_\mathrm{F} \quad . \tag{1.68}$$

The quantity on the right-hand side is called the fluxoid in the loop. With $q_\mathrm{s} = 2e$ it must therefore be equal to an integral multiple of the flux quantum

$$\Phi_0 = \frac{h}{2e} \simeq 2.07 \times 10^{-15} \mathrm{Wb} \quad . \tag{1.69}$$

As we know, the current falls off exponentially in the interior of a superconductor. If the integration path only penetrates far enough into the superconductor, the current term in (1.68) makes no further contribution. Then the flux Φ_F through the loop surface must satisfy the quantisation condition. This is usually the case in practice. The current term comes into the reckoning only in extreme situations, e. g. with a very small opening and an integration path at the inner edge of the opening.

1.6 Effect of Geometry and Magnetic Field

With large currents and strong magnetic fields further effects occur in superconductors. As these usually have only parasitic influence at most in electronic circuits, with their normally small currents and weak magnetic fields, we shall here describe these effects qualitatively rather than quantitatively.

In Sect. 1.3 it has already been pointed out that the superconducting current density is limited by the breaking of the Cooper pairs. This critical volume current density is [1.1]

$$J_c = \frac{e n_s \Delta}{\sqrt{2 m W_F}} \tag{1.70}$$

with the charge e, the density n_s and the mass m of the electrons in the superconducting state, together with the Fermi energy W_F and the energy gap Δ.

Since the current density decreases exponentially in the interior of the superconductor with the penetration depth λ, J_c is first reached on the surface of the superconductor. At that moment there is on the superconductor a magnetic field with strength

$$H = H_c = J_c \lambda \quad . \tag{1.71}$$

This equation follows from (1.9) and the surface current density

$$J_A = \int_0^\infty J \, dx = J_c \int_0^\infty e^{-x/\lambda} dx \tag{1.72}$$

with the coordinates of Fig. 1.7. H_c from (1.71) is known as the thermodynamic critical magnetic field of the superconductor. From J_c according to (1.70) and the geometry of the superconductor there follows the critical current I_c, and hence the maximal conduction current which can be carried in the superconducting state when no external magnetic field is present. In general then a superconductor loses its resistance-free state if somewhere on the surface the magnetic field strength exceeds the critical field strength H_c. The magnetic field strength is then determined not only by the conduction current but also by the screening current induced by an external magnetic field. This is a generalised form of the Silsbee hypothesis [1.1, 16].

These considerations imply that the critical current for a round wire with radius $R \gg \lambda$ is

$$I_c = 2\pi R H_c \quad . \tag{1.73}$$

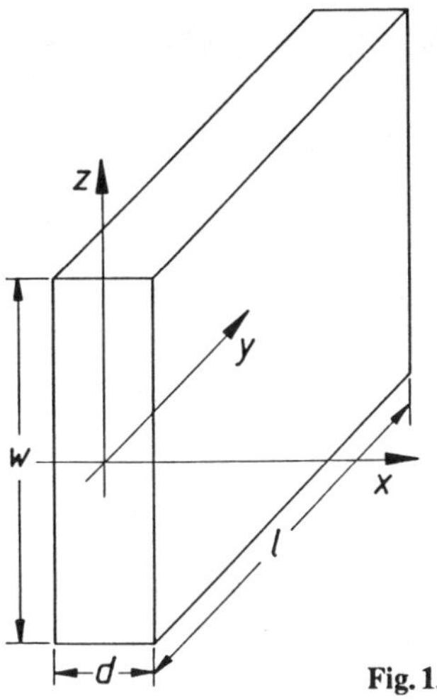

Fig. 1.18. Superconducting plate, $d \ll w \ll l$

For a plate as shown in Fig. 1.18 with a conduction current in the y-direction, solution of the London equations for $l \gg w$ but arbitrary d/λ gives

$$I_c = 2wH_c \tanh(d/2\lambda) \quad . \tag{1.74}$$

With this current the current density at the surface is just equal to $J_c = H_c/\lambda$ [1.1]. According to (1.74) I_c for small thickness d is reduced by the factor $\tanh(d/2\lambda)$ compared with thick plates.

So far we have accounted for the exclusion of the magnetic flux in the superconductor (Meissner effect, see Sect. 1.1) by postulating a screening current inside the penetration depth λ below the surface. We have regarded the superconductor itself as nonmagnetic ($\mu = \mu_0$). This is indeed valid. The effect of the flux exclusion from a superconductor can however also be validly described by another model. Here one starts from the idea that the whole volume of the superconductor has a magnetisation $\boldsymbol{M}$ with

$$\boldsymbol{B} = \mu_0(\boldsymbol{H} + \boldsymbol{M}) \quad , \tag{1.75}$$

where in homogeneous isotropic media $\boldsymbol{H}$ and $\boldsymbol{M}$ are related by the magnetic susceptibility χ_m [1.6]

$$\boldsymbol{M} = \chi_m \boldsymbol{H} \quad , \tag{1.76}$$

so that

$$\boldsymbol{B} = \mu_0(1 + \chi_m)\boldsymbol{H} = \mu_0\mu_r\boldsymbol{H} \quad . \tag{1.77}$$

Since $\boldsymbol{B} = 0$ in the interior of a superconductor, we must have $\chi_m = -1$, $\mu_r = 0$ and $\boldsymbol{M} = -\boldsymbol{H}$. So this second model supposes perfect diamagnetism.

This second model can now first of all be applied in order to calculate the critical magnetic field of superconductors with geometries whose demagnetisation factors are known [1.6]. Thus for the superconducting plate in Fig. 1.18 with a magnetic field in the plane of the plate, i.e. in the y- or z-direction, we have approximately

$$H_{\text{cpar}} = H_c \left(1 - \frac{2\lambda}{d} \tanh \frac{d}{2\lambda}\right)^{-1/2} \quad . \tag{1.78}$$

If d decreases to the order of magnitude of λ, magnetic flux permeates the plate. According to (1.78) the critical magnetic field H_{cpar} then increases. For the extreme case of very thin plates or films with $d \ll 2\lambda$ we even get from (1.78)

$$H_{\text{cpar}} =\simeq \sqrt{3}\frac{2\lambda}{d} H_c \quad . \tag{1.79}$$

For the case of thin plates with perpendicular magnetisation, i.e. in the x-direction of Fig. 1.18, the critical magnetic field decreases drastically compared with H_c. With external magnetic fields $H_e \geq H_c d/w$ the superconductor goes into the so-called intermediate state, in which many superconducting and normally conducting regions exist side by side. For more general geometries this intermediate state is energetically favourable and therefore probable, if [1.1]

$$H_c(1 - n_M) < H_e < H_c \quad . \tag{1.80}$$

Here n_M is the demagnetisation factor, which is dependent only on the geometry [1.6].

Ignoring (1.70) these calculations of the critical currents and magnetic fields of superconductors are based on the London theory. Here it is assumed that the penetration depth, and hence from (1.16) also the density of the superconducting charge carriers, are independent of the magnetic field and of the sample geometry.

The Ginzburg-Landau theory [1.17], see also [1.3], takes account of these effects. Here the superconducting electrons, of density n_s, are described by a wave function ψ with $|\psi|^2 \sim n_s$. It is then assumed that the free energy of the superconducting state deviates from that of the normally conducting state by a difference which can be written as a power series development in $|\psi|^2$. In the neighbourhood of the critical temperature it is sufficient to take account of only two terms of this series development. To the free energy two contributions have to be added. The first is the kinetic energy of the position dependence of ψ, the second, just as in (1.64), that of the presence of a magnetic field, described by the vector potential $\boldsymbol{A}$. The central problem of the Ginzburg-Landau theory is then to find functions $\psi(x, y, z)$ and $\boldsymbol{A}(x, y, z)$, which make the total free energy of a sample the minimum under the relevant boundary conditions. For weak magnetic fields the problem is easily soluble and leads to the London equations. In strong magnetic fields there are only numerical solutions. For the case of a very thick plate with a magnetic field parallel to the surface we find $n_s \sim |\psi|^2$ is constant in the interior of the plate, but decreasing towards the surface. With higher magnetic fields this decrease is more strongly pronounced. Since the penetration depth λ of the magnetic flux density depends as in the London theory on n_s, so also λ depends on the magnetic field. For the case of a thin film, because of the boundary

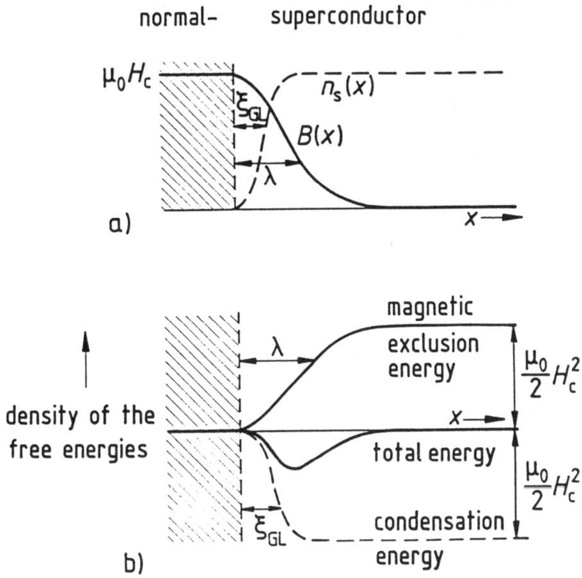

Fig. 1.19a,b. Boundary region between normally conducting and superconducting phases inside a homogeneous material at temperature T. **(a)** Spacewise variation of the magnetic flux density $B(x)$ and the density $n_s(x)$ of superconducting charge carriers. **(b)** Spacewise variation of the free energies

conditions, n_s and therefore also λ depend on the film thickness. So the elements lacking in the London theory are made available by the Ginzburg-Landau theory.

With the aid of results of the Ginzburg-Landau theory we can now describe the magnetic field behaviour of a special class of superconductors, the superconductors of the second type. As the magnetic field is increased these do not go over directly from the Meissner phase to the normally conducting phase, but enter first into the so-called mixed state. Figure 1.19a illustrates this mixed state by the behaviour, calculated from the Ginzburg-Landau theory, of the magnetic flux density $B(x)$ and the density of the superconducting charge carriers $n_s(x)$ in a boundary layer between normal conductor and superconductor. It is here assumed that in the boundary layer just the flux density $B = \mu_0 H_c$ is applied. In addition to the penetration depth λ for the flux density we also show a characteristic length ξ_{GL}, which is a measure of the shortest possible distance over which n_s can change appreciably. ξ_{GL} is known as the Ginzburg-Landau coherence length.

Figure 1.19b shows the behaviour of the energy densities with which $B(x)$ and $n_s(x)$ are connected. The exclusion energy is the energy density connected with the exclusion of the flux from the superconductor, namely $B^2/(2\mu_0)$. The condensation energy is the energy per unit volume which is liberated when normally conducting electrons transform into the superconducting state, see Sect. 1.3. According to whether the Ginzburg-Landau parameter

$$\kappa = \frac{\lambda}{\xi_{GL}} , \tag{1.81}$$

and hence the ratio of the characteristic lengths is small or large, the total energy of the transition between normal- and superconductors can be positive or

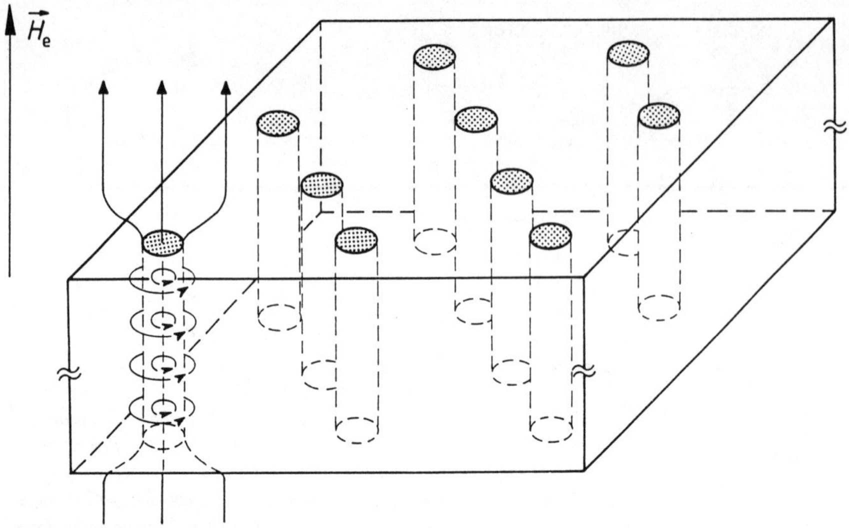

Fig. 1.20. Schematic representation of the mixed state. Magnetic field and superconducting screening currents are indicated only for one flux line. From [1.2]

negative. For large κ and hence for negative total energy such boundary regions become energetically favourable. Figure 1.20 shows how, under these circumstances, many normally conducting cylinders form in the superconductor. They are formed from a part of the external magnetic flux in the shape of flux lines. The diameter of the flux lines is of the order of magnitude of the coherence length ξ_{GL}. Round their superconducting shells flow circular screening currents. This state is called the mixed state, or the Shubnikov phase. The sample becomes entirely normally conducting only with a magnetic field strength H_{c2} in the boundary plane

$$H_{c2} = \sqrt{2}\kappa H_c \quad . \tag{1.82}$$

Below a magnetic field strength of H_{c1} there is no magnetic field even in this superconductor. It is then in the Meissner phase. For $\kappa \gg 1/\sqrt{2}$ we have

$$H_{c1} \simeq \frac{1/2\kappa}{\ln(\kappa + 0.08)H_c} \quad . \tag{1.83}$$

Superconductors which can go into the mixed state are known as superconductors of the second type. Those which are superconducting only in the Meissner phase are called superconductors of the first type. The two types are distinguished by different ratios of λ to ξ_{GL}. From the Ginzburg-Landau theory we find that

$$\text{Superconductor of the first type:} \quad \kappa < 1/\sqrt{2} \tag{1.84}$$

$$\text{Superconductor of the second type:} \quad \kappa > 1/\sqrt{2} \tag{1.85}$$

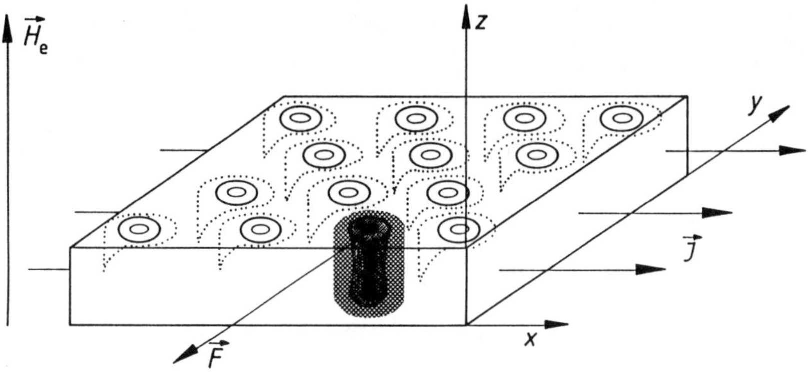

Fig. 1.21. Mixed state with imposed current density J and Lorentz force $\boldsymbol{F}$. The magnetic field distribution in the flux line is shown by blackening. From [1.2]

When a plate in the mixed state, see Fig. 1.21, has a current flowing through it parallel to the plane of the plate, the current is distributed over the whole cross-section of the plate. It is therefore not totally restricted to a thin surface layer. Between the magnetic field of the flux lines and the current density $\boldsymbol{J}$ there is now the Lorentz force $\boldsymbol{F}$, which is proportional to $\boldsymbol{J}$ [1.6], perpendicular to both of them. In completely homogeneous material the flux lines can move perpendicularly to their axes. Since the current through the boundary of the plate is fixed, the flux lines will wander in the direction of $\boldsymbol{F}$. The movement of the flux lines through the superconductor leads to losses, i.e. electrical energy is converted into heat. This energy can only come from the imposed current. An electric potential difference arises in the sample. It acquires an electrical resistance. The critical current of this ideal homogeneous superconductor in the mixed state is therefore zero.

In practical materials, however, there are energetically favourable and hence preferred locations for the flux lines. Such pinning centres can arise, for example, at faults in the crystal lattice, at grain boundaries and at microscopic normally conducting inclusions of an alloy. Such pinning centres are often introduced quite deliberately into superconductors of the second type. In these so-called hard superconductors the flux lines may not move as long as the Lorentz forces are smaller than the restrictive forces, and then the sample has no electrical resistance. The critical current in the mixed state is then not equal to zero. Materials with very high critical currents in the mixed state are used for wires in superconducting magnets. For a detailed description of magnetic flux structures in superconductors see [1.19].

In superconducting electronic circuits pinning centres have a rather parasitic character. In thin films it may be possible for the flux lines to move between them as guiding points. The flux lines can be thus activated both thermally and by varying the current and the magnetic field. As the motion of the flux lines changes so does the distribution of the magnetic field in the sample and round about it. In particular Josephson junctions react with great sensitivity to changes in the magnetic field, see Chaps. 3 and 4.

2. SIS Junctions

Figure 2.1 shows a sandwich structure with the sequence of layers superconductor – insulator – superconductor. Even when $U \neq 0\,\mathrm{V}$ no current will normally flow between the superconducting electrodes. If the insulator is very thin, however, a current can begin to flow because of the quantum mechanical tunnelling effect. For this the thickness should amount at most to a few nanometres. The tunnel current then consists in general of two parts. The first arises from the tunnelling of quasi-particles and the second from the tunnelling of Cooper pairs. This second term leads to the so-called Josephson effects, which we shall discuss in Chaps. 3 and 4. Here we shall first concern ourselves with the tunnelling of quasi-particles and assume that the Cooper pair tunnel current is suppressed by some appropriate means, such as a strong magnetic field.

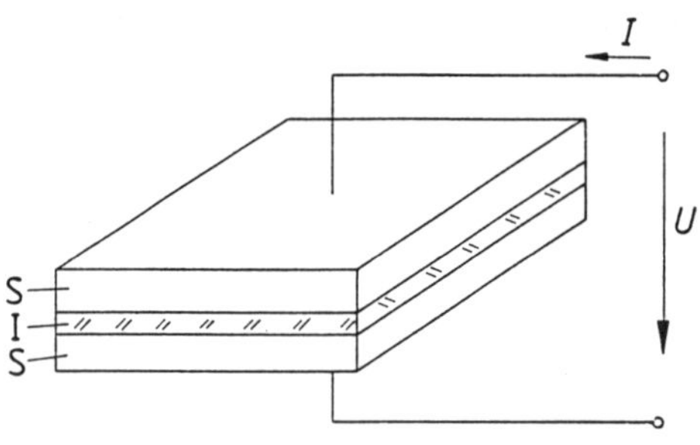

Fig. 2.1. Basic construction of a superconducting tunnel junction. (S) superconductor; (I) insulator, a few nanometres thick

An "SIS junction" primarily means the structure of Fig. 2.1. In a more restricted sense, however, it also means a junction of this type of construction which exploits the quasi-particle tunnelling, and not the Cooper pair tunnelling.

SIS junctions are suitable for detectors and mixers in the millimetre wave region when specially high sensitivity is required. In radio-astronomy SIS mixers are routinely employed.

From the current-voltage characteristics of SIS junctions one can infer the energy gap Δ of the electrode materials [2.1].

2.1 Current-Voltage Characteristics

The band model explained in Sect. 1.3 for the quasi-particles in the superconductor enables one to calculate in broad fashion the current-voltage characteristics of a tunnel junction with a superconducting electrode. The other electrode may consist of a semiconductor, a normal conductor or a superconductor with arbitrary energy gap [2.2]. Here we shall restrict ourselves to the case of two superconductors with equal energy gaps. Figure 2.2 shows the corresponding band model at a temperature $T > 0$. Because of the positive voltage of the right-hand superconductor its energy diagram is lower than that of the left-hand one. In principle, tunnelling can take place at any energy. It is only necessary that electrons in one electrode see allowed and free states at their energy in the other electrode.

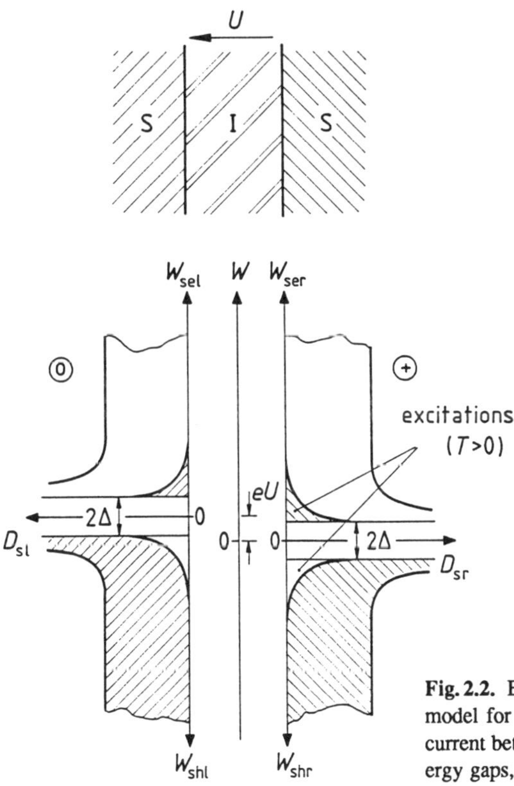

Fig. 2.2. Electrode arrangement and energy band model for determining the quasi-particle tunnelling current between two superconductors with equal energy gaps, $T > 0$, (U) applied voltage

At the temperature $T = 0$ the system is in the ground state: no excitations exist. The electrons of the left-hand superconductor then see in the right-hand superconductor unoccupied states with equal energy, as soon as the applied voltage $U > 2\Delta/e$. As illustrated in Fig. 2.3 for $T = 0$, a current then first begins to flow. Because of the high density of states at the band edge this current starts suddenly, just at the energy gap voltage $U_{gap} = 2\Delta/e$. Although in this thought

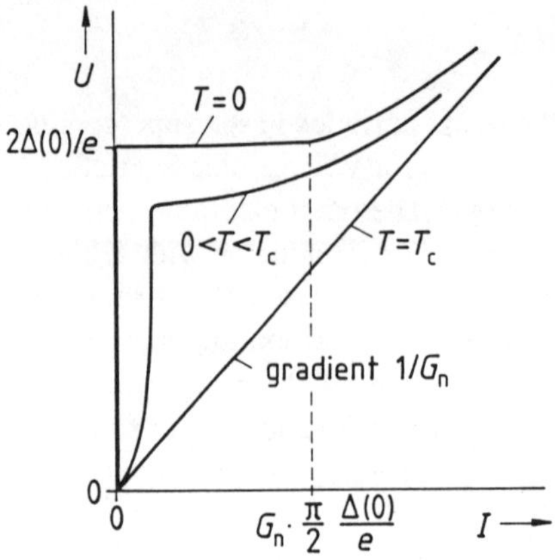

Fig. 2.3. Current-voltage characteristic of the SIS tunnel junction

experiment we have regarded the voltage as the independent variable, in Fig. 2.3 the voltage is plotted against the current, since SIS junctions are often in practice operated with an impressed direct current. In the literature, however, the voltage axis is quite often chosen as the horizontal coordinate axis.

When $T > 0$ the probability of excited or occupied states is described by the Fermi function:

$$f(W) = \frac{1}{1 + \exp(W/k_B T)} \quad . \tag{2.1}$$

To calculate the tunnel current it is now convenient to refer the energies of the electrons and holes in the left-hand and in the right-hand superconductors to the single energy variable W as shown in Fig. 2.2. The total current I is carried by electron currents F_{erl} from the right to the left and F_{elr} from left to right. F_{erl} per energy element dW is proportional to the number of the electrons in the right-hand superconductor and to the number of holes of this energy in the left-hand superconductor. The number of electrons per dW in the right-hand superconductor is equal to the density of states $D_{sr}(W)$ multiplied by the probability $f(W)$ that these states are occupied. In the left-hand superconductor the number of holes per energy element dW is equal to the density of states $D_{sl}(W - eU)$ multiplied by the probability $[1 - f(W - eU)]$ that these states are not occupied. Integration over all energies then gives the electron current from right to left as

$$F_{erl} = \text{const} \int_{-\infty}^{\infty} D_{sr}(W) f(W) D_{sl}(W - eU)[1 - f(W - eU)] dW \quad . \tag{2.2}$$

The electron current from left to right is obtained in a similar manner

$$F_{elr} = \text{const} \int_{-\infty}^{\infty} D_{sl}(W - eU) f(W - eU) D_{sr}(W)[1 - f(W)] dW \quad . \tag{2.3}$$

The total current $I = -e(F_{elr} - F_{elr})$ now follows with $D_{sr} = D_{sl} = D_s$ as

$$I = e \cdot \text{const} \int_{-\infty}^{\infty} D_s(W - eU) D_s(W)[f(W - eU) - f(W)] \mathrm{d}W \quad . \tag{2.4}$$

Here D_s is replaced by (1.41) or (1.42). More detailed consideration shows that the constant appearing in (2.2) to (2.4) is linked as follows

$$\text{const} = \frac{G_n}{e^2 D_n^2(W_n = 0)} \tag{2.5}$$

to the conductance G_n shown by this tunnel junction when its electrodes are in the normally conducting state.

The result of the integration (2.4) for $0 < T < T_c$ is presented in Fig. 2.3. For $T \ll \Delta/k_B$ and $eU < 2\Delta$ the curved branch below the gap voltage $2\Delta/e$ can be represented to a good approximation by the following expression:

$$\begin{aligned} I = & \frac{2G_n}{e} \exp\left(-\frac{\Delta}{k_B T}\right) \sqrt{\frac{2\Delta}{eU + 2\Delta}} (eU + \Delta) \\ & \times \sinh\left(\frac{eU}{2k_B T}\right) K_0 \left(\frac{eU}{2k_B T}\right) \quad . \end{aligned} \tag{2.6}$$

Here K_0 is the modified Bessel function of zero order. The knee appearing in the U, I curve in Fig. 2.3 for $0 \le T < T_c$ is extremely sharp. This pronounced non-linearity is responsible for the outstanding properties of SIS detectors and mixers. To understand these, however, we must first investigate how an SIS tunnel junction behaves when the d.c. voltage U_0 applied to it is superposed with an a.c. voltage of high frequency ω:

$$u(t) = U_0 + \hat{U}_1 \cos \omega t \quad . \tag{2.7}$$

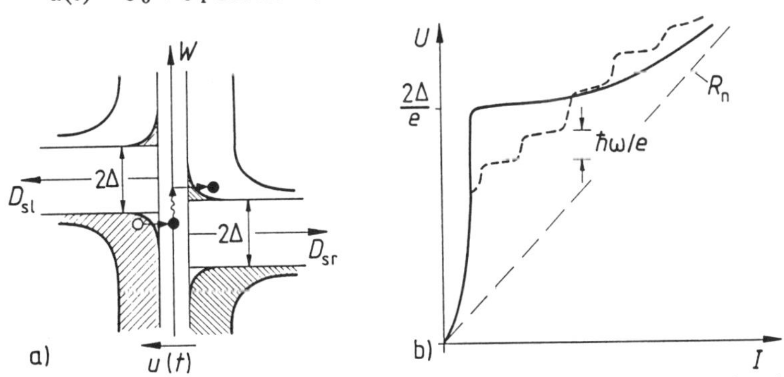

Fig. 2.4a,b. Photon assisted tunnelling process in an SIS junction. (**a**) Energy band model. (**b**) Current-voltage curves. (—) without RF irradiation, (- - -) with RF irradiation

As shown in Fig. 2.4a, a photon assisted tunnelling process then occurs. The electrons can overcome the energy barrier even when the d.c. voltage is still below $2\Delta/e$, if they take up one or more photons of energy $\hbar\omega$. When the increasing d.c. voltage reaches the values $U_0 = 2\Delta/e - n\hbar\omega$, sudden increases in current are to be expected. These jumps in the current are shown in Fig. 2.4b as experimentally observed. For d.c. voltages $U > 2\Delta/e$ similar jumps occur. They are related to the output of photons with energy $n\hbar\omega$.

In order to calculate the currents and voltages we start from the fundamental equation of the quantum mechanical system represented by the right-hand superconductor:

$$\frac{\partial\psi_i(x,t)}{\partial t} = -\frac{\mathrm{j}}{\hbar} W_i \psi_i(x,t) \quad . \tag{2.8}$$

Here, ψ_i are the wave functions and W_i the energies of the system, x a position coordinate and t the time. If we now assume that the right-hand electrode attains a voltage $u(t)$ above the left-hand electrode, the electron energy on the right changes from W_i to $W_i + eu(t)$. The solution of the differential equation (2.8) is then

$$\psi_i(x,t) = \psi_i(x) \exp\left[-\frac{\mathrm{j}}{\hbar}\int_0^t [W_i + eu(t')]dt'\right] \quad . \tag{2.9}$$

If $u(t)$ is now substituted from (2.7), the $\psi_i(x,t)$ can be expressed by the Bessel functions J_n:

$$\psi_i(x,t) = \psi_i(x) \exp\left[-\mathrm{j}(W_i + eU_0)t/\hbar\right] \sum_{n=-\infty}^{\infty} J_n\left(\frac{e\hat{U}_1}{\hbar\omega}\right) \mathrm{e}^{-\mathrm{j}n\omega t} \quad . \tag{2.10}$$

This representation by Bessel functions applies in similar fashion to the sidebands of a frequency modulated high frequency carrier. Equation (2.10) is now capable of the interpretation that a state whose energy differs by $n\hbar\omega$ from $W_i + eU_0$ occurs with a probability $J_n^2(e\hat{U}_1/\hbar\omega)$. This interpretation of (2.10) is schematically indicated in Fig. 2.5. Since all energy states of individual electrons

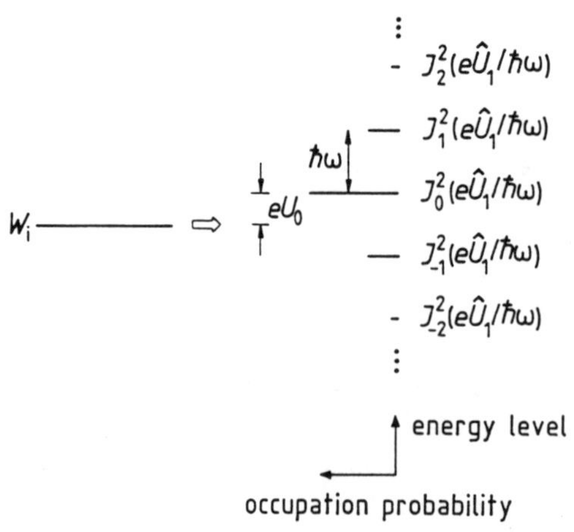

Fig. 2.5. Virtual energy levels of the right-hand electrode from Fig. 2.4, caused by harmonic modulation according to (2.10)

are modulated in the same manner, these virtual shifts in energy level are equivalent to shifts of the d.c. voltage to $U_0 + n\hbar\omega/e$. The probability $J_n^2(e\hat{U}_1/\hbar\omega)$ depends upon the amplitude of the a.c. voltage $\hat{U}_1$.

The resulting d.c. portion of the tunnel current is given by

$$I_0(U_0, \hat{U}_1) = \sum_{n=-\infty}^{\infty} J_n^2\left(\frac{e\hat{U}_1}{\hbar\omega}\right) I_{\text{dc}}\left(U_0 + \frac{n\hbar\omega}{e}\right) \quad . \tag{2.11}$$

Here $I_{\text{dc}}(U_0)$ denotes the unmodulated I, U characteristic. The a.c. voltage $\hat{U}_1 \cos\omega t$ accordingly displaces the voltage of the unmodulated I, U characteristic by integral multiples of $\hbar\omega/e$. The amplitudes of the individual current increments are given by $J_n^2(e\hat{U}_1/\hbar\omega)$.

The in-phase component $\hat{I}_{1W}$ of the alternating current can be obtained from the following equation:

$$\frac{\hat{I}_{1W}\hat{U}_1}{2\hbar\omega} = \sum_{n=-\infty}^{\infty} \frac{n}{e} J_n^2\left(\frac{e\hat{U}_1}{\hbar\omega}\right) I_{\text{dc}}\left(U_0 + \frac{n\hbar\omega}{e}\right) \quad . \tag{2.12}$$

The left-hand side of (2.12) represents the total rate of photons which are taken up, the right-hand side is the sum of the photon rates for the tunnelling processes at the energy levels n. These rates are found from the product of the number n of photons which are needed to overcome the energy threshold, multiplied by the appropriate rate of electrons produced. This is equal to the portion of the direct current caused by tunnelling into the energy level n, divided by the elementary charge. From (2.12) it follows by a recursion formula for Bessel functions that

$$\begin{aligned} \hat{I}_{1W}(U_0, \hat{U}_1) = \sum_{n=-\infty}^{\infty} J_n\left(\frac{e\hat{U}_1}{\hbar\omega}\right) \left[J_{n+1}\left(\frac{e\hat{U}_1}{\hbar\omega}\right) + J_{n-1}\left(\frac{e\hat{U}_1}{\hbar\omega}\right)\right] \\ \times I_{dc}\left(U_0 + \frac{n\hbar\omega}{e}\right) \quad . \end{aligned} \tag{2.13}$$

Equation (2.13) can also be derived more rigorously, see [2.3].

Receivers with the highest sensitivity are constructed from SIS junctions. Their intrinsic noise is therefore of special interest. It can be represented by a noise current source run in parallel with a noise-free SIS junction. For the spectral density function, giving the mean square current for each Hertz of band width, we have in general for this quasi-particle tunnel junction [2.4]

$$\begin{aligned} S_{iqp}(f_{\text{N}}) = e\left\{ I_{\text{dc}}\left(U_0 + \frac{hf_{\text{N}}}{e}\right) \coth\left[\frac{1}{k_{\text{B}}T}\left(\frac{eU_0 + hf_{\text{N}}}{2}\right)\right] \right. \\ \left. + I_{\text{dc}}\left(U_0 - \frac{hf_{\text{N}}}{e}\right) \coth\left[\frac{1}{k_{\text{B}}T}\frac{(eU_0 - hf_{\text{N}})}{2}\right]\right\} \quad . \end{aligned} \tag{2.14}$$

Here f_{N} is the noise frequency. For practical applications (2.14) is variously evaluated, according to whether the voltage U_0 at the point of application is small or large compared with hf_N/e at the noise frequency under consideration. For $U_0 \gg hf_N/e$ we have, for example,

$$S_{iqp} = 2eI_{dc}(U_0)\coth\frac{eU_0}{2k_BT} \quad . \tag{2.15}$$

And again, according to whether the energy eU_0 taken from the electrons in passing through the barrier is small or large compared with the thermal energy k_BT, (2.15) reduces to the well known formulae for the classical thermal and shot noise, respectively:

$$S_{iqp} = 4k_BT\frac{I_{dc}(U_0)}{U_0} = \frac{4k_BT}{R} \quad \text{for} \quad U_0 \ll k_BT/e \tag{2.16}$$

$$S_{iqp} = 2eI_{dc}(U_0) \quad \text{for} \quad U_0 \gg k_BT/e \quad . \tag{2.17}$$

The right-hand side of (2.16) assumes a linear current-voltage characteristic with resistance R. When $T = 4.2\,\mathrm{K}$, then $k_BT/e = 362\mu\mathrm{V}$.

2.2 Detectors

Using the nonlinear current-voltage characteristics of the SIS junction, RF oscillations can be directly rectified. In addition to the classical rectification, caused by the curvature of the characteristic graph, quantum effects also arise. Of particular practical importance is the rectification of very weak RF signals, which due to the lack of suitable RF amplifiers cannot be amplified before rectification. Limits to the sensitivity of response are set by the intrinsic noise of the SIS junctions. We shall here first of all calculate the current sensitivity of the SIS detector and then its noise equivalent power NEP. Finally we shall give an example.

For a fixed working point voltage U_0, at small amplitude $\hat{U}_1$, of the RF oscillation, the difference between the direct current with and without RF input is given by (2.11) as

$$\Delta I_{dc} = \frac{\hat{U}_1^2}{4}\,\frac{I_{dc}(U_0+\hbar\omega/e) - 2I_{dc}(U_0) + I_{dc}(U_0 - \hbar\omega/e)}{(\hbar\omega/e)^2} \quad . \tag{2.18}$$

The amplitude of the output current is proportional to the square of the amplitude of the RF voltage. It thus amounts to a quadratic detector. The current sensitivity R_i is defined as the ratio of the detector current ΔI_{dc} to the received RF power. This RF power is calculated as half the product of $\hat{U}_1$ and $\hat{I}_{1W}$ from (2.13). For small amplitudes $\hat{U}_1$ (2.13) simplifies to

$$\hat{I}_{1W} \simeq \hat{U}_1\frac{I_{dc}(U_0+\hbar\omega/e) - I_{dc}(U_0-\hbar\omega/e)}{2\hbar\omega/e} \quad . \tag{2.19}$$

The current sensitivity then follows as

$$R_i = \frac{\Delta I_{dc}}{\hat{U}_1 \hat{I}_1/2} = \frac{e}{\hbar\omega} \frac{I_{dc}(U_0 + \hbar\omega/e) - 2I_{dc}(U_0) + I_{dc}(U_0 - \hbar\omega/e)}{I_{dc}(U_0 + \hbar\omega/e) - I_{dc}(U_0 - \hbar\omega/e)} \quad . \tag{2.20}$$

If the current-voltage curve $I_{dc}(U_0)$ varies smoothly over the voltage $\hbar\omega/e$ on each side of the working point, then R_i reduces to the classical limiting case:

$$R_i \simeq \frac{1}{2} \frac{d^2 I_{dc}/dU_0^2}{dI_{dc}/dU_0}, \quad \text{classical limit} \quad . \tag{2.21}$$

This classical result would imply that the detector would become arbitrarily sensitive as the curvature $|d^2 I_{dc}/dU_0^2|$ of the direct current curve kept on increasing. However, (2.20) shows that there is a fundamental limit: for a working point at the knee of the continuous curve of Fig. 2.4b, where the curvature is sharp, $I_{dc}(U_0 - \hbar\omega/e) \simeq I_{dc}(U_0)$ whereas $I_{dc}(U_0 + \hbar\omega/e) \gg I_{dc}(U_0)$. Thereupon (2.20) becomes

$$R_i \simeq \frac{e}{\hbar\omega}, \quad \text{quantum limit} \quad . \tag{2.22}$$

In the region of this quantum limit, therefore, each absorbed photon causes one electron to tunnel through the energy barrier.

The noise of the SIS detector at ordinary working points is determined by the shot noise of (2.17), the flicker noise with the noise power density given by $P_f' = K_f/f_N$ where K_f is a constant independent of the frequency, and the noise power density $k_B T_e$ from the following amplifier. The amplifier is characterised by its equivalent noise temperature T_e. In what follows R_d will denote the effective internal resistance of the SIS junction at low frequencies. If the amplifier input impedance is matched to R_d the noise power present at the amplifier input becomes

$$P_{N\,av} = \frac{R_d}{2} e I_0 B + K_f \ln\frac{f_2}{f_1} + k_B T_e B \quad . \tag{2.23}$$

Here the frequencies f_1 and f_2 with $f_2 - f_1 = B$ are the cutoff frequencies of a band-pass filter placed before the display. It will now be assumed that the SIS junction takes up a radio frequency power P_{RF} which is rectangularly modulated in amplitude. The available rectified power in the fundamental oscillation of the rectangular modulation is then

$$P_{Det\,av} = \frac{1}{\pi^2} \Delta I_{dc}^2 R_d = \frac{1}{\pi^2} R_d R_i^2 P_{RF}^2 \quad . \tag{2.24}$$

The minimal detectable RF power is now obtained when (2.23) and (2.24) are equal. It follows that

$$P_{RF\,min} = \frac{\pi}{R_i} \sqrt{\frac{e I_0 B}{2} + \frac{K_f}{R_d} \ln\frac{f_2}{F_1} + \frac{k_B T_e B}{R_d}} \quad . \tag{2.25}$$

$P_{\mathrm{RF\,min}}$ in (2.25) is about $\frac{\pi}{2}$ greater than the noise equivalent power NEP, which is also quoted as the criterion.

For the current I_0 determining the shot noise we have strictly speaking

$$I_0 = I_{\mathrm{dc}}(U_0) + \Delta I_{\mathrm{dc}} \quad . \tag{2.26}$$

The "dark current" $I_{\mathrm{dc}}(U_0)$, however, is usually much greater than the direct current ΔI_{dc}.

The flicker noise plays an important part only at low modulation frequencies. If it is neglected and also the intrinsic noise of the amplifier vanishes ($T_{\mathrm{e}} = 0$), it follows from (2.25) at the quantum limit that

$$P_{\mathrm{RF\,min}} = \frac{\pi}{\sqrt{2}}\hbar\omega\sqrt{\frac{I_0 B}{e}} \quad . \tag{2.27}$$

In the calculation of $P_{\mathrm{RF\,min}}$ it was assumed that the impedances were matched at the input. The real part of the input admittance is given as $\hat{I}_{1W}/\hat{U}_1$ from (2.19). The imaginary part, which is determined essentially by the parallel plate capacity of the SIS junction, is compensated by a tuning circuit if necessary.

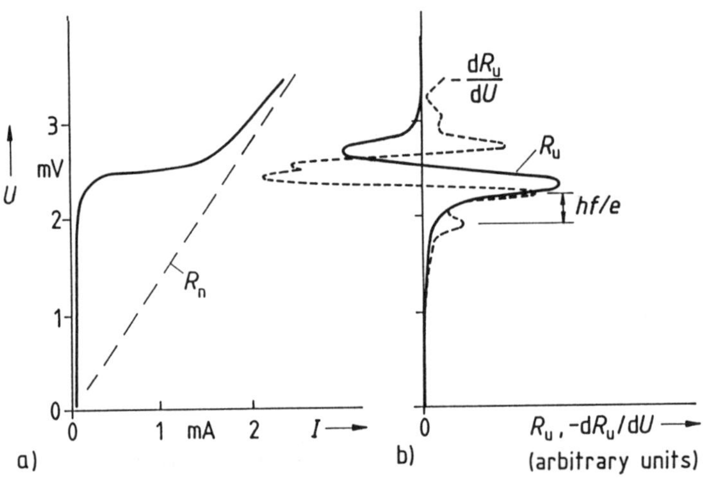

Fig. 2.6. SIS detector according to [2.5]. **(a)** Current-voltage curve. **(b)** Voltage sensitivity $R_u = R_i R_d$ together with $-\mathrm{d}R_u/\mathrm{d}U$ as functions of the voltage U at the working point

Experiments with SIS detectors are described, for example, in [2.5]. SIS junctions were used, made in thin film technology from crossed strips with the stratification Pb-In/oxide/Pb. With an area of $12\,\mu\mathrm{m}^2$ the capacity of the junction amounted to 0.36 pF. At 4.2 K, at the knee of the $U(I)$ curve, see Fig. 2.6, $(\mathrm{d}^2I/\mathrm{d}U^2)/(\mathrm{d}I/\mathrm{d}U) \simeq 20\,000\,\mathrm{V}^{-1} \gg 8\,000\,\mathrm{V}^{-1} \simeq e/\hbar\omega$ for $\omega/2\pi = 70\,\mathrm{GHz}$, so that this is not a classical but a quantum detector. The normal resistance amounts to $R_{\mathrm{n}} \simeq 93\,\Omega$. The junction with its glass substrate is mounted into a rectangular waveguide of reduced height, see Fig. 2.7. The susceptance at the reception frequency is compensated by a waveguide sliding short circuit. There remains an input resistance of about $R_n/3$.

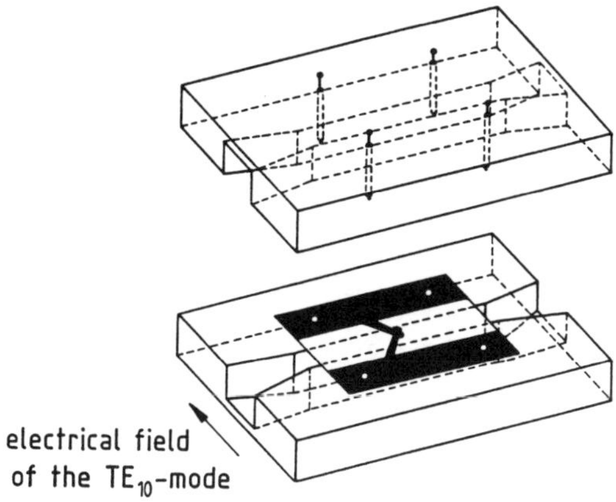

Fig. 2.7. Rectangular waveguide of reduced height with inserted SIS junction on glass substrate

Figure 2.6 shows also the voltage sensitivity $R_u = R_i R_d$ of the detector as a function of the direct voltage at the working point. Since this curve was obtained at unusually high RF power, it already shows a "multi-photon" structure. It is especially clear in the differentiated curve.

With weak signals at 70 GHz the measured current sensitivity R_i is found to bc a factor 0.46 of the quantum limit $c/\hbar\omega$ = 3450 A/W from (2.22). The noise equivalent power NEP was measured for the case of a rectangular modulation with 413 Hz (chopper frequency). It gave NEP = 1.7×10^{-15}W in 1 Hz display band width. Mismatch at the RF input, flicker noise still effective at 413 Hz and the amplifier noise temperature T_e = 80 K are presumably the reasons for not quite achieving the limit of the noise equivalent power of $2.1 \times 10^{-16}\mathrm{W}/\sqrt{\mathrm{Hz}}$ implied by (2.27) with $I_0 = I_{dc}(U_0) = 1.7\,\mu$A. The detector was saturated by an input power of about 10^{-9}W.

2.3 SIS Mixers

A detailed theoretical description of the SIS mixer is to be found in [2.3]. We shall here consider only the most important aspects needed for the understanding and design of an SIS mixer. Before we look at the SIS mixer itself, however, we shall in the next section explain a few fundamental points in the calculation of the conversion gain of a mixer with a nonlinear circuit element.

2.3.1 Conversion Matrix and Gain of a Mixer

According to [2.6] one can describe the frequency conversion at a nonlinear circuit element by means of a conversion matrix, which connects the small signal currents and voltages at all frequencies arising in the element. For a mixer with voltage control as in Fig. 2.8 the conversion matrix is an admittance matrix $\boldsymbol{Y}$

$$\underline{\boldsymbol{I}} = \boldsymbol{Y}\underline{\boldsymbol{U}} \quad . \tag{2.28}$$

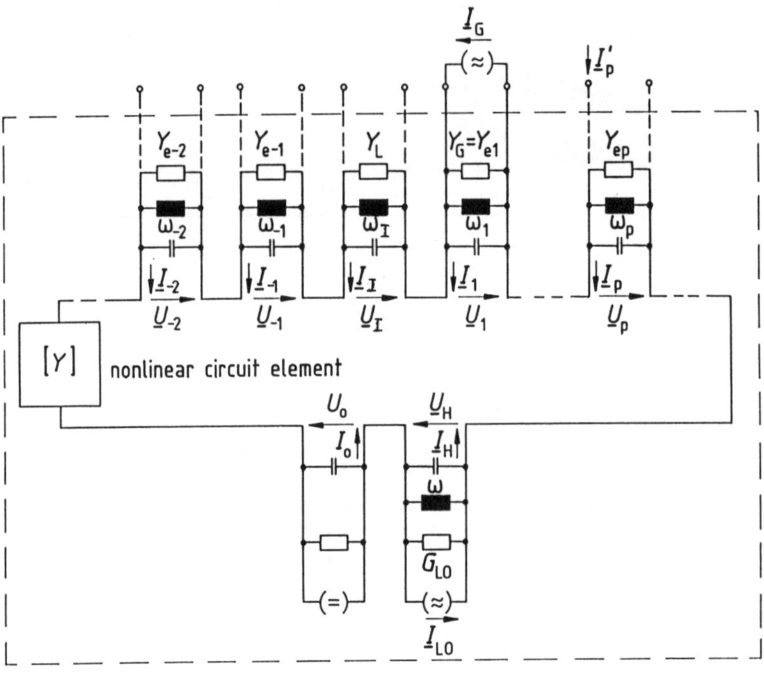

Fig. 2.8. Equivalent circuit of a mixer with voltage control

$\underline{\boldsymbol{I}}$ and $\underline{\boldsymbol{U}}$ are column vectors whose elements are the phasors of the small signal currents $\underline{I}_p$ and voltages $\underline{U}_p$ at the frequencies $\omega_p = \omega_I + p\omega$ with $p = 0, \pm 1, \pm 2, \ldots$. Here ω is the frequency of the local oscillator and ω_I the intermediate frequency. In Fig. 2.8 the index $p = 0$ is substituted by the letter I in order to make it clear that we are here dealing with the intermediate frequency. With fundamental mode mixers as normally used $\omega_S = \omega_1$ is the frequency of the incident signal and $\omega_B = |\omega_{-1}|$ the image frequency. The auxiliary or local oscillator (LO) oscillation $\underline{U}_H, \underline{I}_H$ with the frequency ω has a very much greater amplitude than the small signals. Direct voltage U_0 or direct current I_0 determine together with $|U_H|$ or $|I_H|$ the operating point of the mixer. We have

$$\omega_S = \omega + \omega_I \tag{2.29}$$

$$\omega_B = \omega - \omega_I = 2\omega - \omega_S \quad . \tag{2.30}$$

At the small signal frequencies ω_p the mixer is terminated with the admittance Y_{ep} by the external circuit. Here $Y_{el} = Y_G$ is the internal admittance of the generator with the short circuit current $\underline{I}_G$ at the frequency ω_1. $Y_L = Y_I$ represents the load admittance at the intermediate frequency ω_I. First of all we shall consider how the conversion gain of the mixer, that means the ratio of the output power to the available input power

$$G = |\underline{U}_I|^2 \mathrm{Re}(Y_L) \frac{4\mathrm{Re}(Y_G)}{|\underline{I}_G|^2} \tag{2.31}$$

depends on the elements of the conversion matrix $\boldsymbol{Y}$ and the elements of the external circuit. Following [2.7] we collect these elements into a diagonal matrix:

$$\mathrm{diag}(Y_{ep}) = \begin{bmatrix} & & \vdots & & \\ & Y_1 & 0 & 0 & \\ \cdots & 0 & Y_I & 0 & \cdots \\ & 0 & 0 & Y_{-1} & \\ & & \vdots & & \end{bmatrix} \quad . \tag{2.32}$$

Furthermore we put all small signal current sources (normally this is only the generator at the signal frequency ω_1) outside the dashed boundary in Fig. 2.8. For the column vector $\underline{\boldsymbol{I}}'$, whose elements are the small signal current sources, we then have

$$\underline{\boldsymbol{I}}' = \{\mathrm{diag}(Y_{ep}) + Y\}\underline{\boldsymbol{U}} \quad . \tag{2.33}$$

For the voltage vector $\underline{\boldsymbol{U}}$ we then get

$$\underline{\boldsymbol{U}} = \boldsymbol{Z}'\underline{\boldsymbol{I}}' \tag{2.34}$$

with

$$\boldsymbol{Z}' = \boldsymbol{Y}'^{-1} = \{\mathrm{diag}(Y_{ep}) + \boldsymbol{Y}\}^{-1} \quad . \tag{2.35}$$

If the only small signal current source is at the input, (2.34) can be rewritten as

$$\begin{bmatrix} \vdots \\ \underline{U}_1 \\ \underline{U}_I \\ \underline{U}_{-1} \\ \vdots \end{bmatrix} = \begin{bmatrix} & & \vdots & & \\ & Z'_{11} & Z'_{1I} & Z'_{1-1} & \\ \cdots & Z'_{I1} & Z'_{II} & Z'_{I-1} & \cdots \\ & Z'_{-11} & Z'_{-1I} & Z'_{-1-1} & \\ & & \vdots & & \end{bmatrix} \begin{bmatrix} \vdots \\ \underline{I}_G \\ 0 \\ 0 \\ \vdots \end{bmatrix} \quad . \tag{2.36}$$

We also have $\underline{U}_I = Z'_{I1}\underline{I}_G$. It then follows from (2.31) that the conversion gain is

$$G = 4|Z'_{I1}|^2 \mathrm{Re}(Y_G)\mathrm{Re}(Y_L) \quad . \tag{2.37}$$

One usually wishes to design the mixer so that G is a maximum. To show how to achieve this we shall study the simple example of a single side band mixer with $Y_{-1} \to \infty$ and $Y_{\text{ep}} \to \infty$ when $|p| > 1$. The external circuit will thus short circuit all currents with frequencies ω_{-1} and ω_{p} for $|p| > 1$. We shall furthermore assume for the sake of simplicity that the elements of the conversion matrix $\boldsymbol{Y}$ together with the internal admittance of the generator and load are real. Accordingly, if the index 1 is replaced by S for "Signal"

$$\boldsymbol{Y} = \begin{bmatrix} G_{\text{SS}} & G_{\text{SI}} \\ G_{\text{IS}} & G_{\text{II}} \end{bmatrix} \tag{2.38}$$

$$\text{diag}(Y_{\text{ep}}) = \begin{bmatrix} G_{\text{G}} & 0 \\ 0 & G_{\text{L}} \end{bmatrix} \quad . \tag{2.39}$$

From (2.35) the matrix $\boldsymbol{Z}'$ now becomes

$$\begin{aligned} \boldsymbol{Z}' &= \begin{bmatrix} G_{\text{SS}} + G_{\text{G}} & G_{\text{SI}} \\ G_{\text{IS}} & G_{\text{II}} + G_{\text{L}} \end{bmatrix}^{-1} \\ &= \frac{1}{(G_{\text{SS}} + G_{\text{G}})(G_{\text{II}} + G_{\text{L}}) - G_{\text{SI}} G_{\text{IS}}} \begin{bmatrix} G_{\text{II}} + G_{\text{L}} & -G_{\text{SI}} \\ -G_{\text{IS}} & G_{\text{SS}} + G_{\text{G}} \end{bmatrix} \quad . \end{aligned} \tag{2.40}$$

Using (2.37) we now find the gain to be

$$G = \frac{4 G_{\text{IS}}^2 G_{\text{G}} G_{\text{L}}}{[(G_{\text{SS}} + G_{\text{G}})(G_{\text{II}} + G_{\text{L}}) - G_{\text{SI}} G_{\text{IS}}]^2} \quad . \tag{2.41}$$

When the load conductance $G_{\text{L}} \to 0$ and also when $G_{\text{L}} \to \infty$ the gain vanishes; in between there is a maximum. And there is a similar dependence on the generator conductance G_{G}. Setting to zero the derivatives of (2.41) with respect to G_{L} and G_{G} yields optimal values of G_{L} and G_{G} for maximal power gain G_{m}, namely

$$G_{\text{L}} = G_{\text{II}} \sqrt{1 - \eta} \tag{2.42}$$

$$G_{\text{G}} = G_{\text{SS}} \sqrt{1 - \eta} \tag{2.43}$$

$$G_{\text{m}} = \frac{G_{\text{IS}}}{G_{\text{SI}}} \frac{\eta}{\left(1 - \sqrt{1 - \eta}\right)^2} \tag{2.44}$$

with

$$\eta = \frac{G_{\text{SI}} G_{\text{IS}}}{G_{\text{II}} G_{\text{SS}}} \quad . \tag{2.45}$$

For classical mixers $G_{\text{IS}} = G_{\text{SI}}$ and $0 \leq \eta \leq 1$. As η increases the maximal gain also increases from 0 to 1.

In the following section we shall study the peculiarities of mixers in the quantum regime.

We have so far assumed that we know the elements of the conversion matrix $\boldsymbol{Y}$ in (2.28). But how does one find them? In the case of the control of the nonlinear element by the auxiliary oscillation, when the instantaneous differential conductance is known, the elements of $\boldsymbol{Y}$ can be determined as coefficients of a Fourier series, in which the differential conductance is developed as a function of time.

In the case when the instantaneous differential conductance is not known, one can still determine the conversion matrix under certain circumstances from the direct current characteristic and the large signal input admittance Y_{H} at the auxiliary frequency ω [2.8]. The functions

$$I_0 = I_0(U_0, |\underline{U}_{\mathrm{H}}|) \tag{2.46}$$

$$\underline{I}_{\mathrm{H}} = \underline{U}_{\mathrm{H}} Y_{\mathrm{H}}(U_0, |\underline{U}_{\mathrm{H}}|) \tag{2.47}$$

must be known either from experiments or from theory.

We shall now proceed to explain how one determines $\boldsymbol{Y}$ in this second case. Since $|\underline{U}_{\mathrm{H}}| = \sqrt{\underline{U}_{\mathrm{H}}\underline{U}_{\mathrm{H}}^*}$ we can consider I_0 and $\underline{I}_{\mathrm{H}}$ in (2.46) and (2.47) as functions of $U_0, \underline{U}_{\mathrm{H}}$ and $\underline{U}_{\mathrm{H}}^*$. For their total differentials we have

$$\mathrm{d}I_0 = \frac{\partial I_0}{\partial U_0}\mathrm{d}U_0 + \frac{1}{2}\frac{\underline{U}_{\mathrm{H}}^*}{|\underline{U}_{\mathrm{H}}|}\frac{\partial I_0}{\partial |\underline{U}_{\mathrm{H}}|}\mathrm{d}\underline{U}_{\mathrm{H}} + \frac{1}{2}\frac{\underline{U}_{\mathrm{H}}}{|\underline{U}_{\mathrm{H}}|}\frac{\partial I_0}{\partial |\underline{U}_{\mathrm{H}}|}\mathrm{d}\underline{U}_{\mathrm{H}}^* \tag{2.48}$$

$$\begin{aligned}\mathrm{d}\underline{I}_{\mathrm{H}} =& \underline{U}_{\mathrm{H}}\frac{\partial Y_{\mathrm{H}}}{\partial U_0}\mathrm{d}U_0 + \left(Y_{\mathrm{H}} + \frac{|\underline{U}_{\mathrm{H}}|}{2}\frac{\partial Y_{\mathrm{H}}}{\partial |\underline{U}_{\mathrm{H}}|}\right)\mathrm{d}\underline{U}_{\mathrm{H}} \\ &+ \frac{1}{2}\frac{\underline{U}_{\mathrm{H}}^2}{|\underline{U}_{\mathrm{H}}|}\frac{\partial Y_{\mathrm{H}}}{\partial |\underline{U}_{\mathrm{H}}|}\mathrm{d}\underline{U}_{\mathrm{H}}^* \quad .\end{aligned} \tag{2.49}$$

In what follows the index 1 will be replaced by S for signal frequency and -1 by B for image frequency. It will now be assumed that all currents $\underline{I}_{\mathrm{p}}$ with $|p| > 1$ are short-circuited at the nonlinear element, and hence $\underline{U}_{\mathrm{p}} = 0$ for $|p| > 1$. The voltages with frequencies $\omega_{\mathrm{S}} = \omega + \omega_{\mathrm{I}}$ and $\omega_{\mathrm{B}} = \omega - \omega_{\mathrm{I}}$ are regarded as small perturbations on the auxiliary voltage $\underline{U}_{\mathrm{H}}$, so that the timewise variation of the voltage at the RF input can be written as

$$\mathrm{Re}\left[(\underline{U}_{\mathrm{H}} + \mathrm{d}\underline{U}_{\mathrm{H}})\mathrm{e}^{\mathrm{j}\omega t}\right] = \mathrm{Re}\left(\underline{U}_{\mathrm{H}}\mathrm{e}^{\mathrm{j}\omega t}\right) + \mathrm{Re}\left(\mathrm{d}\underline{U}_{\mathrm{H}}\mathrm{e}^{\mathrm{j}\omega t}\right) \tag{2.50}$$

with

$$\mathrm{d}\underline{U}_{\mathrm{H}} = \underline{U}_{\mathrm{S}}\mathrm{e}^{\mathrm{j}\omega_{\mathrm{I}} t} + \underline{U}_{\mathrm{B}}\mathrm{e}^{-\mathrm{j}\omega_{\mathrm{I}} t} \tag{2.51}$$

where $|\underline{U}_{\mathrm{S}}|, |\underline{U}_{\mathrm{B}}| \ll |\underline{U}_{\mathrm{H}}|$ and $\omega_{\mathrm{I}} \ll \omega$. Then the intermediate frequency voltage can also be regarded as a small perturbation on the direct voltage:

$$U_0 \to U_0 + \mathrm{d}U_0 \quad \text{with} \quad \mathrm{d}U_0 = \mathrm{Re}\left(\sqrt{2}\underline{U}_{\mathrm{I}}\mathrm{e}^{\mathrm{j}\omega_{\mathrm{I}} t}\right) \quad . \tag{2.52}$$

The $\sqrt{2}$ here takes account of the fact that $|\underline{U}_{\mathrm{I}}|$ represents a root-mean-square value. As in (2.51) and (2.52) the currents are now similarly replaced by

$$\mathrm{d}\underline{I}_1 = \underline{I}_{\mathrm{S}}\mathrm{e}^{\mathrm{j}\omega_{\mathrm{I}} t} + \underline{I}_{\mathrm{B}}\mathrm{e}^{-\mathrm{j}\omega_{\mathrm{I}} t} \tag{2.53}$$

$$\mathrm{d}I_0 = \mathrm{Re}\left(\sqrt{2}\underline{I}_{\mathrm{I}}\mathrm{e}^{\mathrm{j}\omega_{\mathrm{I}}t}\right) \quad . \tag{2.54}$$

Now (2.51) to (2.54) are substituted in (2.48) and (2.49), and sorted according to the different time factors. We then have

$$\underline{I}_{\mathrm{S}} = \left(Y_{\mathrm{H}} + \frac{|\underline{U}_{\mathrm{H}}|}{2}\frac{\partial Y_{\mathrm{H}}}{\partial|\underline{U}_{\mathrm{H}}|}\right)\underline{U}_{\mathrm{S}} + \frac{\underline{U}_{\mathrm{H}}}{\sqrt{2}}\frac{\partial Y_{\mathrm{H}}}{\partial U_0}\underline{U}_{\mathrm{I}} + \frac{\underline{U}_{\mathrm{H}}^2}{2|\underline{U}_{\mathrm{H}}|}\frac{\partial Y_{\mathrm{H}}}{\partial|\underline{U}_{\mathrm{H}}|}\underline{U}_{\mathrm{B}}^* \tag{2.55}$$

$$\underline{I}_{\mathrm{I}} = \frac{\underline{U}_{\mathrm{H}}^*}{\sqrt{2}|\underline{U}_{\mathrm{H}}|}\frac{\partial I_0}{\partial|\underline{U}_{\mathrm{H}}|}\underline{U}_{\mathrm{S}} + \frac{\partial I_0}{\partial U_0}\underline{U}_{\mathrm{I}} + \frac{\underline{U}_{\mathrm{H}}}{\sqrt{2}|\underline{U}_{\mathrm{H}}|}\frac{\partial I_0}{\partial|\underline{U}_{\mathrm{H}}|}\underline{U}_{\mathrm{B}}^* \tag{2.56}$$

$$\underline{I}_{\mathrm{B}}^* = \frac{\underline{U}_{\mathrm{H}}^{*2}}{2|\underline{U}_{\mathrm{H}}|}\frac{\partial Y_{\mathrm{H}}^*}{\partial|\underline{U}_{\mathrm{H}}|}\underline{U}_{\mathrm{S}} + \frac{\underline{U}_{\mathrm{H}}^*}{\sqrt{2}}\frac{\partial Y_{\mathrm{H}}^*}{\partial U_0}\underline{U}_{\mathrm{I}} + \left(Y_{\mathrm{H}}^* + \frac{|\underline{U}_{\mathrm{H}}|}{2}\frac{\partial Y_{\mathrm{H}}^*}{\partial|\underline{U}_{\mathrm{H}}|}\right)\underline{U}_{\mathrm{B}}^* \quad . \tag{2.57}$$

If we write the conversion matrix in the following form

$$\begin{bmatrix}\underline{I}_{\mathrm{S}}\\ \underline{I}_{\mathrm{I}}\\ \underline{I}_{\mathrm{B}}^*\end{bmatrix} = \begin{bmatrix}Y_{\mathrm{SS}} & Y_{\mathrm{SI}} & Y_{\mathrm{SB}}\\ Y_{\mathrm{IS}} & Y_{\mathrm{II}} & Y_{\mathrm{IB}}\\ Y_{\mathrm{BS}} & Y_{\mathrm{BI}} & Y_{\mathrm{BB}}\end{bmatrix}\begin{bmatrix}\underline{U}_{\mathrm{S}}\\ \underline{U}_{\mathrm{I}}\\ \underline{U}_{\mathrm{B}}^*\end{bmatrix} \tag{2.58}$$

$$\underline{\boldsymbol{I}} = \boldsymbol{Y}\underline{\boldsymbol{U}} \quad , \tag{2.59}$$

we can assign to the elements of $\boldsymbol{Y}$ their values by comparison with (2.55) to (2.57). Without loss of generality we can also choose the time zero so that $\underline{U}_{\mathrm{H}}$ is real and we can then replace $\underline{U}_{\mathrm{H}}$ by the real root-mean-square value of the time function U_{H}. This gives the following results:

$$Y_{\mathrm{SS}} = Y_{\mathrm{H}} + \frac{U_{\mathrm{H}}}{2}\frac{\partial Y_{\mathrm{H}}}{\partial U_{\mathrm{H}}} \tag{2.60}$$

$$Y_{\mathrm{SI}} = \frac{U_{\mathrm{H}}}{\sqrt{2}}\frac{\partial Y_{\mathrm{H}}}{\partial U_0} \tag{2.61}$$

$$Y_{\mathrm{SB}} = \frac{U_{\mathrm{H}}}{2}\frac{\partial Y_{\mathrm{H}}}{\partial U_{\mathrm{H}}} \tag{2.62}$$

$$Y_{\mathrm{IS}} = \frac{1}{\sqrt{2}}\frac{\partial I_0}{\partial U_{\mathrm{H}}} \tag{2.63}$$

$$Y_{\mathrm{II}} = \frac{\partial I_0}{\partial U_0} \tag{2.64}$$

$$Y_{\mathrm{IB}} = Y_{\mathrm{IS}}^* \,; \quad Y_{\mathrm{BS}} = Y_{\mathrm{SB}}^* \,; \quad Y_{\mathrm{BI}} = Y_{\mathrm{SI}}^* \,; \quad Y_{\mathrm{BB}} = Y_{\mathrm{SS}}^* \quad . \tag{2.65}$$

From (2.60) to (2.65) the elements of the conversion matrices for the SIS mixer and the Josephson mixer can be calculated.

2.3.2 Conversion Gain of the SIS Mixer

We shall now proceed to calculate the conversion gain of the SIS mixer. As in Sect. 2.3.1 we shall here assume that the external circuit allows voltage drops across the SIS junction for only a finite number of frequencies. All other current components will be short-circuited.

A quantum theoretical calculation of the conversion matrix shows that its elements are in general complex [2.3]. The imaginary parts, however, are small, and in what follows they will be completely neglected.

In order to find the simplest possible representation which however still demonstrates the essentials, we shall moreover only allow for voltages at the frequency of the incoming signal ω_S, at the intermediate frequency ω_I and at the frequency $\omega = \omega_S - \omega_I$ of the local oscillator. All other frequencies, and in particular the image frequency $\omega - \omega_I = 2\omega - \omega_S$, will accordingly be short-circuited by the external circuit. The conversion matrix indicates the relationship between the phasors of the currents and the voltages at the signal and intermediate frequencies:

$$\begin{bmatrix} \underline{I}_S \\ \underline{I}_I \end{bmatrix} = \begin{bmatrix} G_{SS} & G_{SI} \\ G_{IS} & G_{II} \end{bmatrix} \begin{bmatrix} \underline{U}_S \\ \underline{U}_I \end{bmatrix} \quad . \tag{2.66}$$

All four elements G_{ij} in general differ from one another.

We cannot calculate the G_{ij} immediately as Fourier coefficients [2.6], since there is no expression for the differential resistance at the SIS junction which is valid at high frequencies (ω_S, ω) in the same way as at lower frequencies, see Sect. 2.3.1. We can, however, calculate the G_{ij} from (2.60), (2.61), (2.63) and (2.64). Since $\hat{U}_1 = \sqrt{2}|\underline{U}_H| = \sqrt{2}\underline{U}_H$ the functional form of the direct current characteristic given in (2.11) can be modified as in (2.46) to become

$$I_0(U_0, U_H) = \sum_{n=-\infty}^{\infty} J_n^2(\alpha) I_{dc}(U_0 + n\hbar\omega/e) \tag{2.67}$$

with

$$\alpha = \frac{\sqrt{2}eU_H}{\hbar\omega} \quad . \tag{2.68}$$

Moreover, from (2.13) with $\hat{I}_{1W} = \sqrt{2}\underline{I}_H = \sqrt{2}I_H$, we find that

$$I_H(U_0, U_H) = U_H Y_H(U_0, U_H) \quad , \tag{2.69}$$

where

$$\begin{aligned} Y_H(U_0, U_H) = &\frac{1}{U_H} \sum_{n=-\infty}^{\infty} J_n(\alpha)[J_{n+1}(\alpha) + J_{n-1}(\alpha)] \\ &\times I_{dc}\left(U_0 + \frac{n\hbar\omega}{e}\right) \quad . \end{aligned} \tag{2.70}$$

Setting $Y_{SS} = G_{SS}$ in (2.60), and using (2.68) and (2.70) together with the addition theorems for Bessel functions, we obtain

$$G_{SS} = \frac{1}{2}\frac{e}{\hbar\omega}\sum_{n=-\infty}^{\infty}\left[J_{n-1}^2(\alpha) - J_{n+1}^2(\alpha)\right] I_{dc}\left(U_0 + \frac{n\hbar\omega}{e}\right) \quad . \tag{2.71}$$

In similar fashion, with $Y_{SI} = G_{SI}, Y_{IS} = G_{IS}$ and $Y_{II} = G_{II}$, (2.61), (2.63) and (2.64), together with (2.67), (2.68) and (2.70), yield

$$G_{SI} = \frac{1}{2}\sum_{n=-\infty}^{\infty} J_n(\alpha)\left[J_{n+1}(\alpha) + J_{n-1}(\alpha)\right]\frac{d}{dU_0} I_{dc}\left(U_0 + \frac{n\hbar\omega}{e}\right) \tag{2.72}$$

$$G_{IS} = \frac{e}{\hbar\omega}\sum_{n=-\infty}^{\infty} J_n(\alpha)\left[J_{n-1}(\alpha) - J_{n+1}(\alpha)\right] I_{dc}\left(U_0 + \frac{n\hbar\omega}{e}\right) \tag{2.73}$$

$$G_{II} = \sum_{n=-\infty}^{\infty} J_n^2(\alpha)\frac{d}{dU_0} I_{dc}\left(U_0 + \frac{n\hbar\omega}{e}\right) \quad . \tag{2.74}$$

We note that the elements G_{SS} and G_{II} as well as G_{SI} and G_{IS} are not equal. According to (2.44) the maximal conversion gain is

$$G_m = \frac{G_{IS}}{G_{SI}}\frac{\eta}{(1+\sqrt{1-\eta})^2} \tag{2.75}$$

with

$$\eta = \frac{G_{IS}G_{SI}}{G_{II}G_{SS}} \quad . \tag{2.76}$$

Here it is assumed that the generator internal conductance G_G is adjusted to the input conductance $G_S = \underline{I}_S/\underline{U}_S$ and the load conductance G_L is adjusted to the output conductance $G_I = \underline{I}_I/\underline{U}_I$, i.e. $G_G = G_S$ and $G_L = G_I$. Generator and load conductances are then according to (2.42) and (2.43)

$$G_G = G_{SS}\sqrt{1-\eta} \tag{2.77}$$

and

$$G_L = G_{II}\sqrt{1-\eta} \quad . \tag{2.78}$$

In order to interpret (2.75) to (2.78), the elements of the conversion matrix are plotted in Fig. 2.9a and the maximal gain G_m together with the parameter η as a function of the normalised amplitude $\hat{U}_H = \sqrt{2}U_H$ of the local oscillator voltage in Fig. 2.9b. The idealised $I_{dc}(U_0)$ curve of Fig. 2.10 lies at the basis of Fig. 2.9. That is indeed a great simplification compared with Fig. 2.3, but the essentials remain recognisable in the results. For the calculation of the curves in Fig. 2.9 a typical operating point was chosen with $U_0 = U_{gap} - \hbar\omega/(2e)$[2.9].

We recognize that here also, as with the classical mixer, the maximal conversion gain is greatest when $\eta = 1$. From Fig. 2.9b, moreover, G_m is then clearly

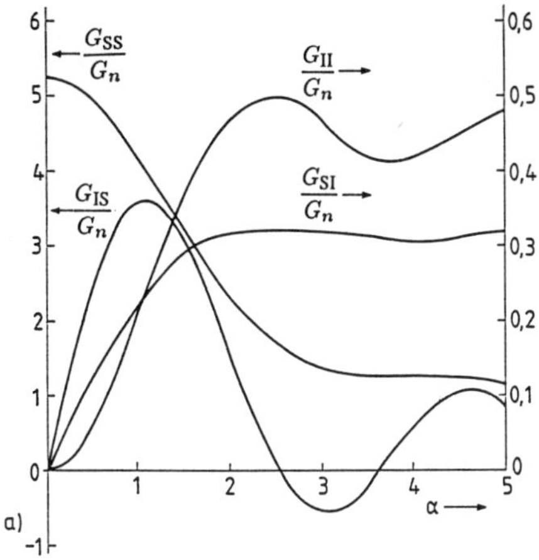

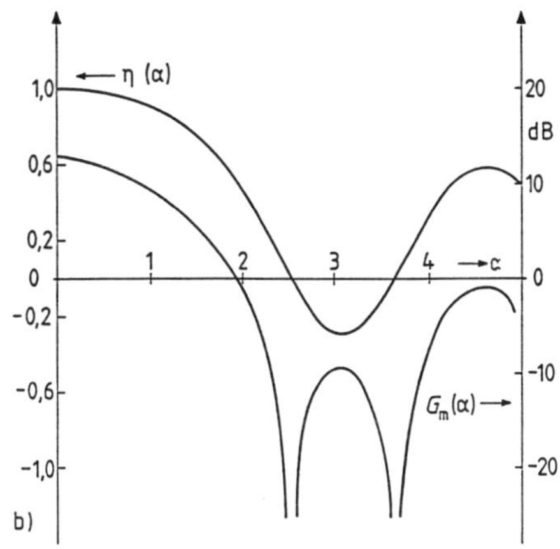

Fig. 2.9a,b. Characteristic quantities of the SIS mixer as a function of the normalised LO amplitude α, based in idealised $I_{dc}(U_0)$ curve from Fig. 2.10, $\hbar\omega/e = 0.1U_{gap}$, operating point at $U_0 = U_{gap} - \hbar\omega/(2e)$. **(a)** Elements of the conversion matrix. **(b)** Parameter η and maximal gain G_m

greater than $1(\hat{=} > 0\,\mathrm{dB})$. According to Fig. 2.9b, however, $\eta = 1$ means that $\alpha = 0$ and the oscillator amplitude $\hat{U}_H$ vanishes. This would appear to contradict the assumption $|\underline{U}_H| \gg |\underline{U}_S|$. In practice, however, the fact is that when $\alpha \to 0$ and $\eta \to 0$ the conductances G_G and G_L vanish for optimal adjustment and hence impedance match is no longer possible. The minimal values of α must accordingly be held between 1 and 2. For example, for a minimal output conductance of $G_I = G_L = 1/1000\,\Omega$ in a typical SIS junction with a normal conductance of $G_n = 1/200\,\Omega$ we find from (2.78) and Fig. 2.9a a minimal value of $\alpha = 1.6$. Then the maximal conversion gain $G_m = 3(\hat{=}\ 4.8\,\mathrm{dB})$ and the input resistance $1/G_S = 120\,\Omega$.

According to Fig. 2.9b, when $1 \le \alpha < 1.9$ a real conversion amplification is still possible. At least in part, however, it is in practice lost again through the imperfect $I_{dc}(U_0)$ characteristic and mismatch.

Since the input resistance for the LO signal is of the order of $1/G_n$ the required LO power is given by (2.86). Assuming $\alpha = 2$ we obtain

$$P_{LO} \simeq 2G_n\left(\frac{\hbar\omega}{e}\right)^2 \quad . \tag{2.79}$$

This power, lying in the region of a few nanowatts to microwatts, is very much smaller than the LO power required for classical mixers.

So far we have assumed that the LO voltage $\hat{U}_H$ is impressed, i.e. it is independent of the operation point or, more generally, of the impedance of the SIS mixer. For this an LO source with an infinitely large internal admittance would

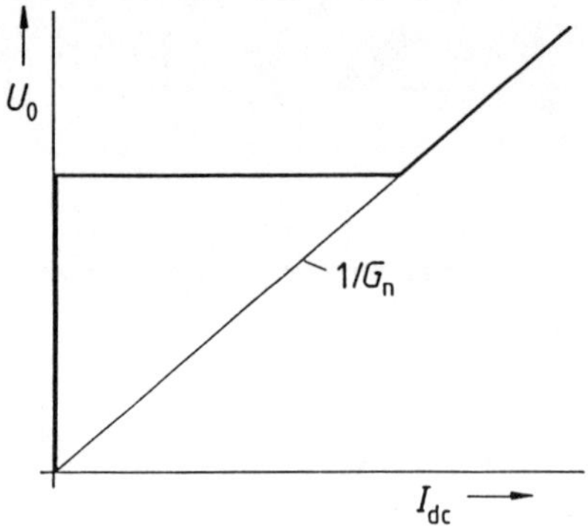

Fig. 2.10. Idealised $I_{dc}(U_0)$ curve

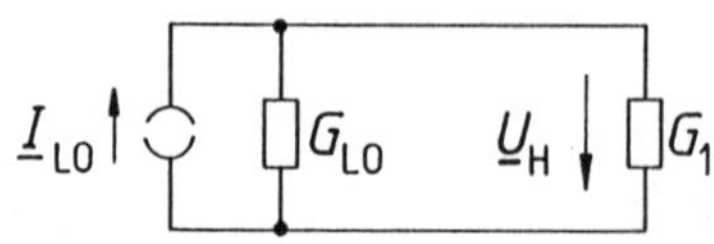

Fig. 2.11. Equivalent circuit of the local oscillator LO with the input conductance G_1 of the SIS mixer at the LO frequency

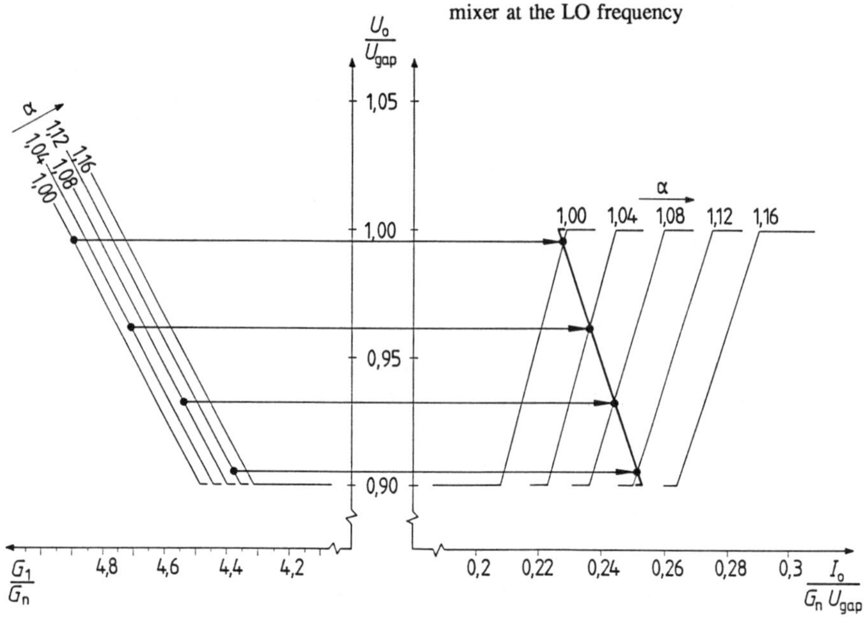

Fig. 2.12. Illustration of the negative differential resistance in the $I_0(U_0)$ characteristic based on Fig. 2.9 and $\hbar\omega/e = 0.1U_{gap}$ with $U_{gap} = 2\Delta/e$; (*left*) input conductance G_1/G_n as a function of the d.c. voltage U_0/U_{gap} with normalised LO amplitude α as parameter, the points correspond to $G_1/G_n = 4.9/\alpha$; (*right*) direct current $I_0/(G_nU_{gap})$ as a function of U_0/U_{gap} with α as parameter. The points on the right are obtained by projection from those of the left of the figure for equal values of U_0/U_{gap} and α

be required. In actual fact the LO source always has a finite internal admittance or conductance, respectively. According to Fig. 2.11 the LO voltage $\hat{U}_H$ then depends on the input conductance G_1 of the mixer at the LO frequency ω. In the left-hand side of Fig. 2.12 $G_1 = \hat{I}_{1W}/\hat{U}_1$ is plotted from (2.13) with $\hat{U}_1 = \hat{U}_H$ as a function of U_0 with $\alpha = e\hat{U}_H/\hbar\omega$ as parameter. The effect examined here, namely that a negative differential resistance occurs in the $U_0(I_0)$ curve, becomes most obvious for vanishing internal conductance $G_{LO} = 0$. Then $\hat{I}_{LO} = \hat{U}_H G_1$, and, using (2.68),

$$\frac{G_1}{G_n} = \frac{e}{\hbar\omega}\frac{\hat{I}_{LO}}{G_n}\frac{1}{\alpha} \quad . \tag{2.80}$$

If the LO short circuit current $\hat{I}_{LO}$ is held constant, then (2.80) represents a condition connecting G_1 and α. For example, if $\hat{I}_{LO} = 4.9 \times \hbar\omega G_n/e$ we get the points plotted on the left-hand side of Fig. 2.12. On the right-hand side we have plotted the d.c. curves $I_0(U_0)$ according to (2.11) with α as parameter. If the points from the left are projected over to the curves on the right with equal α values, we obtain the d.c. curves applicable to the specified impressed LO current. The resulting curve clearly shows negative differential resistance. This is also observable in practice, although not quite so clearly as in Fig. 2.12, since $G_{LO} > 0$.

If the operating point of an SIS mixer lies on a curve like this with a negative differential resistance, then our previous calculations of the maximal conversion gain are no longer valid. The output resistance of the mixer is then negative. This effect can be used for further amplification. However, this does involve stability problems.

For computerised treatment of SIS mixers one should consult [2.10].

2.3.3 Noise of the SIS Mixer

The intrinsic noise of SIS mixers comes close to the lower limit imposed by quantum theory.

The order of magnitude of this limit can be estimated from the Heisenberg uncertainty principle. According to this the smallest possible reception power can be represented by the quotient of the energy $\hbar\omega_S$ of a photon which is taken up, divided by the observation time, the latter being about the reciprocal of the band width B. This power is now equated to the thermal noise power $k_B T_M B$ available in B from a fictitious resistance at the temperature T_M. It follows that the limit for the mixer noise temperature is

$$T_M \simeq \frac{\hbar\omega_S}{k_B} \quad . \tag{2.81}$$

This limit is here formulated in a very general way. For the SIS mixer in particular this can actually be substantiated, at least to the order of magnitude. For this we start again from the single side band mixer with voltage control, in which all the current components apart from those at the signal and intermediate frequencies are short-circuited. Since the noise figure and hence also the noise temperature of a linear two-port are independent of the load impedance, we may even short-circuit also the intermediate frequency components, see Fig. 2.13. The intrinsic noise of the SIS junction is described by a noise current source i_r in parallel. Only its frequency components at f_S and f_I contribute to the noise

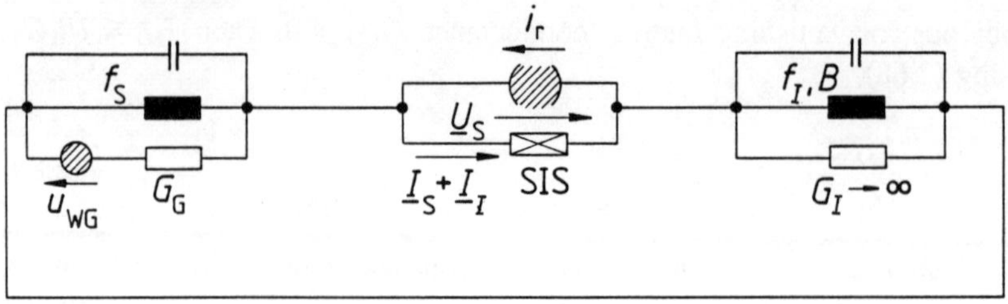

Fig. 2.13. Equivalent circuit for the noise of the single side band SIS mixer

currents at the intermediate frequency output. All other frequency components of i_r are short-circuited.

In contrast to (2.14) we shall make the bold simplification that the noise source i_r has equal noise power densities inside the band width B around f_S and f_I, and that this is given by (2.17), the usual equation for the shot noise. Thus for both frequency components we have

$$\overline{i_r^2} = 2eBI_0(U_0) \quad . \tag{2.82}$$

Here I_0 is the current according to (2.11), controlled according to the curves "pumped" by the LO signal. Both noise components are now replaced for the calculation by sinusoidal currents with equal root-mean-square values:

$$\underline{I}_{rS} = \underline{I}_S = \sqrt{\overline{i_r^2}}\,; \quad \underline{I}_{rI} = \sqrt{\overline{i_r^2}} \quad . \tag{2.83}$$

$|\underline{I}_{rI}^2|$ contributes directly to the short-circuit noise current, $|\underline{I}_{rS}^2|$ only over a frequency conversion. When $G_I \to \infty$, then $\underline{U}_I \to 0$. The conversion matrix (2.66) then becomes

$$\underline{I}_S = G_{SS}\underline{U}_S\,; \quad \underline{I}_I = G_{IS}\underline{U}_S \quad . \tag{2.84}$$

The short-circuited noise current caused by i_r at the intermediate frequency output accordingly appears as

$$|\underline{I}_{rK}|^2 = \overline{i_r^2}\left(1 + \frac{G_{IS}^2}{(G_G + G_{SS})^2}\right) \quad . \tag{2.85}$$

The thermal noise of the generator conductance, which is at a temperature T, is described by the voltage source u_{WG} with $\overline{u_{WG}^2} = 4k_BTB/G_G$. It causes a short-circuited noise current $\underline{I}_{WK}$ at the intermediate frequency output given by

$$|\underline{I}_{WK}|^2 = 4k_BTB\frac{G_{IS}^2 G_G}{(G_G + G_{SS})^2} \quad . \tag{2.86}$$

From this noise figure, F_M and the noise temperature T_M of the mixer can be found according to

$$F_M = 1 + \frac{|\underline{I}_{rK}|^2}{|\underline{I}_{WK}|^2} = 1 + \frac{T_M}{T} \quad . \tag{2.87}$$

With (2.82) T_M becomes

$$T_M = \frac{e}{2k_B} I_0(U_0) \frac{(G_G + G_{SS})^2 + G_{IS}^2}{G_G G_{IS}^2} \quad . \tag{2.88}$$

T_M increases beyond all limits if the generator internal resistance G_G tends to zero or infinity. Minimal noise temperature T_M, i.e. noise match, occurs when

$$G_G = \sqrt{G_{SS}^2 + G_{IS}^2} \tag{2.89}$$

and is then

$$T_{M\,min} = \frac{e}{2k_B} I_0(U_0) \frac{2}{G_{IS}^2} \left[\sqrt{G_{SS}^2 + G_{IS}^2} + G_{SS} \right] \quad . \tag{2.90}$$

Using the idealised $I_{dc}(U_0)$ curve of Fig. 2.10, (2.11), (2.71) and (2.73) can be expanded in terms of small values of the parameter α, about the typical operation point $U_0 = U_{gap} - \hbar\omega/(2e) \simeq U_{gap}$, where the energy gap voltage $U_{gap} = 2\Delta/e$:

$$I_0(U_0, \hat{U}_{LO}) \simeq \frac{\alpha^2}{4} G_n U_{gap} \tag{2.91}$$

$$G_{SS} \simeq G_n \frac{eU_{gap}}{2\hbar\omega} \tag{2.92}$$

$$G_{IS} \simeq G_n \frac{\alpha}{2} \frac{eU_{gap}}{\hbar\omega} \quad . \tag{2.93}$$

The minimal mixer noise temperature (2.90) then becomes

$$T_{M\,min}(\alpha \to 0) = \frac{\hbar\omega}{k_B} \quad . \tag{2.94}$$

If $\omega \simeq \omega_S$ this agrees with the general estimate from (2.81). Other estimations lead to half the value quoted in (2.94), see [2.3].

The quantum limit of the noise temperature for the SIS mixer with short-circuited image frequency was derived here from the shot noise of the SIS junction. The same minimal noise temperature is also found for the double side band SIS mixer with the same termination at signal and image frequencies, if the idealised curves and small LO amplitude are again assumed [2.3].

For the numerical calculation of the noise temperature of actual SIS mixers, more accurately than is possible in this section, refer to [2.11, 19].

2.3.4 Properties of Actual SIS Mixers

In the calculations of the conversion gain in Sect. 2.3.2 the reactive components of the conversion matrix were neglected. Mainly, however, we failed to take account of the parallel capacitance C of the SIS junction (see Fig. 2.1), which is conditioned by the construction. This was justified so long as $\omega R_n C \ll 1$. In practice, however, one usually chooses $\omega R_n C$ somewhat greater than 1. Then the susceptance $\omega_S C$ must indeed be compensated by the external circuit at the signal frequency, but the higher frequency components are still better short-circuited, thus reducing the noise temperature and increasing the conversion gain. The larger is $\omega R_n C$, the narrower is the band width of the resulting parallel resonance circuit and the better can a single side band mixer be realised.

The conversion gain measured on good SIS mixers normally lies between about -2 dB at 30 GHz, -2 dB at 100 GHz [2.12] and -11 dB at 240 GHz. The high output impedance here often contributes to the mismatch losses. A conversion gain with $G > 1$, which is possible according to Sect. 2.3.2, has already, however, been confirmed experimentally: in [2.13,14] a mixer experiment with Sn junctions is described, which at $f_S = 36$ GHz resulted in a single side band noise temperature of $T_{MSSB} = (9 \pm 6)$ K and a conversion gain of $(+4.3 \pm 1)$ dB. In [2.15], moreover, a mixer for $f_S \simeq 100$ GHz was presented, which showed a double side band noise temperature of $T_{MDSB} = (16.4 \pm 1.8)$ K and a conversion gain of $(+2.6 \pm 0.5)$ dB with a load resistance of $R_I = 700\,\Omega$.

According to Fig. 2.9b the conversion gain of SIS mixers depends strongly on α, so considerable attention should be paid in operation to avoid amplitude oscillations of the local oscillator signal [2.16].

The conversion gain of SIS mixers is normally already reduced at really small signal power by saturation effects. Here the intermediate frequency voltage amplitude $\hat{U}_I$ and the step heights $\hbar\omega/e$ are crucial. At the SIS junction the intermediate frequency voltage U_I is regarded as being superposed on the d.c. voltage. U_I sees a differential resistance independent of amplitude only so long as it is small compared with the interval $\hbar\omega/e$, in which the $I_0(U_0, \hat{U}_{LO})$ curves are approximately linear (see Fig. 2.4). If $N > 1$ SIS junctions are connected in series the saturation power P_{sat} is increased, without losing on the gain G and the noise temperature T_M of the mixer [2.17]. P_{sat} is then given by [2.3]

$$P_{sat} = \frac{(\gamma_0 N \hbar\omega)^2}{2e^2 G R_I} \tag{2.95}$$

with $\gamma_0 = U_{Smax} e/(N\hbar\omega)$. A gain compression of 1 dB corresponds to the value $\gamma_0 \simeq 0.20$. Equation (2.95) in relation to the minimal reception power $k_B T_M B$ gives the maximal dynamic range of the SIS mixer. Dynamic ranges of more than 30 dB should be possible.

Figure 2.14 shows the measured noise temperatures for mixer (T_M) and receiver (T_R). Here the receiver noise temperature

$$T_R = T_M + T_e/G \tag{2.96}$$

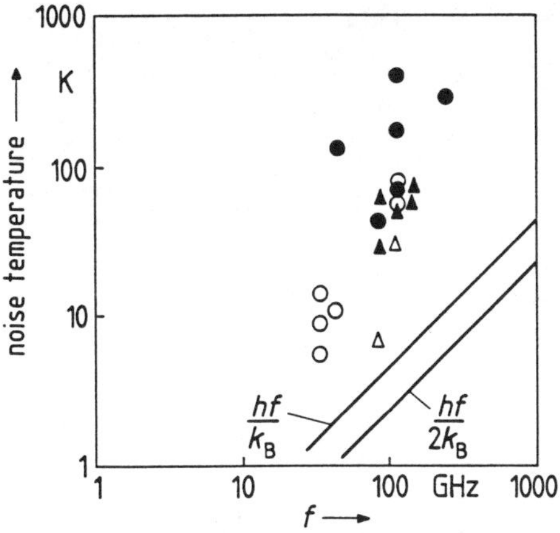

Fig. 2.14. Measured noise temperatures of SIS mixers; (○) single side band and (△) double side band mixer noise temperatures; (●) single side band and (▲) double side band receiver noise temperatures

is influenced quite strongly by the noise temperature T_e of the following intermediate frequency amplifier. For $f_I = 1 \ldots 4$ GHz this is usually a GaAs MESFET or HEMT amplifier cooled to low temperatures with $T_e \simeq 10 \ldots 16$ K or occasionally a parametric amplifier with $T_e \simeq 20$ K. The amplifier noise temperature divided by the mixer gain G contributes to the receiver noise temperature. It is therefore important that the mixer produces as high a gain as possible with $G > 1$. Only then can the mixer noise temperature, which according to Fig. 2.14 approaches the quantum limit, be used to best advantage.

The SIS mixers whose noise temperatures are described by Fig. 2.14 are usually operated at a temperature of $T \simeq 2$ K. It is thereby ensured that the pronounced nonlinearity of the $I_{dc}(U_0)$ curves (Fig. 2.4) is restricted to voltage bands narrower than $\hbar\omega/e$ at frequencies in the millimetre wave region. This, however, also sets the lower frequency limit, up to which the SIS mixers can work in nonclassical mode.

The upper frequency limit is presumably set by additional noise effects and frequency conversion products caused by the Josephson effect (see Chaps. 3 and 4). The Josephson currents involved can indeed be partially suppressed by a magnetic d.c. field, though one nevertheless expects limits, which lie at about 300 GHz for SIS junctions with Pb electrodes. For junctions with high T_c this limit increases proportionally to the energy gap voltage.

Mixers with SIS junctions, in contrast to those with Schottky diodes at the same temperatures, show lower noise temperatures and sometimes a higher conversion gain. The reason for this is that

- the shot noise is much less because of the smaller current at the operating point,
- no additional noise effects occur, such as that caused by the bulk resistance with the Schottky diode,
- the nonlinearity in the direct current curve is much more pronounced.

Finally we show in Fig. 2.15 the construction of an SIS mixer for 90 ... 140 GHz. Figure 2.15a shows the waveguide construction in split block tech-

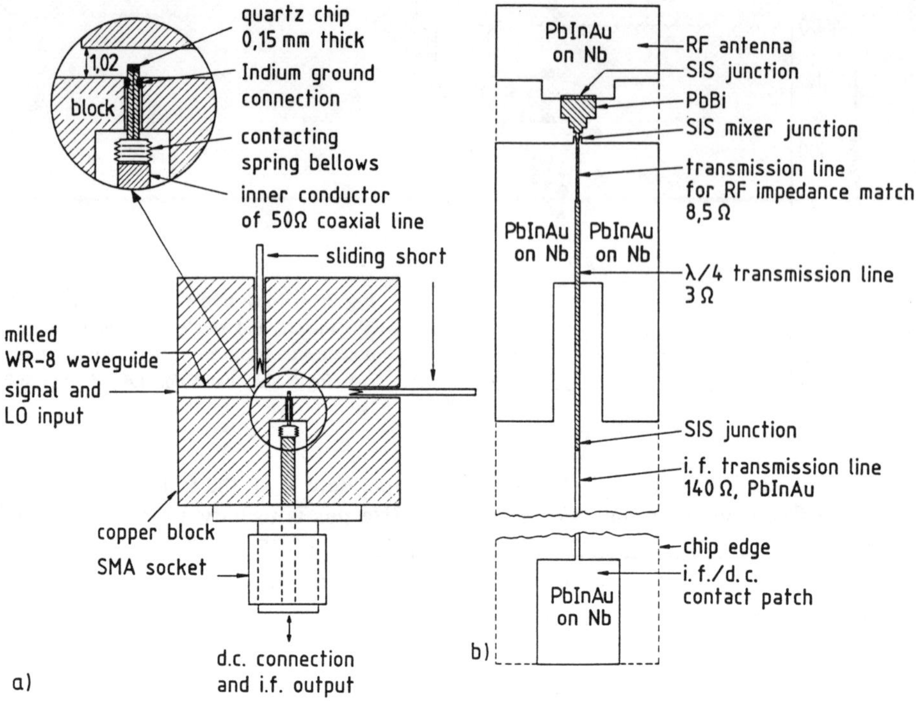

Fig. 2.15a,b. Design of a mixer from [2.18]. **(a)** Mixer block. **(b)** Quartz chip; 0.38 mm × 2.41 mm × 0.15 mm

nique [2.18]. Two sliding shorts are used for the RF impedance match. The chip of 0.15 mm thick crystalline quartz, on which the SIS junction and additional circuit elements are fabricated by thin film technology, projects into the waveguide with one end acting as an antenna. The chip is mounted in a channel in the copper block and, being mechanically decoupled by a spring bellows, is on the output side attached to the inner conductor of a 50 Ω coaxial conductor. The connection between the chip and the gold-plated copper block is made of two pieces of indium pressed between them. The chip itself is shown in Fig. 2.15b. It is covered on one side by an overall base-plate (Nb), an insulating layer (Nb_2O_5), the ground electrode (Pb-In-Au), the window- and insulator-layers (SiO) and the counter electrode (Pb-Bi). The Pb-In-Au ground electrode of the SIS mixer junction is connected to the Nb base-plate. The counter electrode is connected to the antenna by the Pb-Bi metal deposit and a contact formed by a large-area SIS junction. Moreover at the counter electrode there is a short ($< \lambda/4$) relatively high impedance transmission line which is almost short-circuited at the end by the $\lambda/4$ transmission line. By means of this short matching transmission line the capacitance of the SIS junction can be approximately compensated at the signal frequency so that the sliding short has only to compensate the susceptance of the antenna. After the $\lambda/4$ transmission line, via a further SIS junction and the intermediate frequency transmission line, there follows the intermediate frequency and d.c. contact patch.

With this mixer construction, the noise temperature measured at operating temperatures of 3 K was $T_{MDSB} = 20 \ldots 40$ K and the measured conversion gain was 0 dB.

When one takes account of mismatch on the intermediate frequency side, there is a maximal possible gain of $G_m \simeq +3$ dB.

3. Josephson Junctions

Quasi-particles are not the only things that can tunnel through the insulator layer between the two superconducting electrodes of the SIS structure shown in Fig. 2.1: so can Cooper pairs. The associated phenomena are called Josephson effects [3.1] and the components in which these effects are observed and utilised are called Josephson junctions. First of all the a.c. Josephson effect offers interesting applications in the microwave region. We shall describe the basis of this effect, so far as is required in this connection, in Sect. 3.1, and typical properties of Josephson junctions in Sects. 3.2 and 3.3. A first application, namely as a magnetic field detector, follows in Sect. 3.4. More detailed presentations are to be found in [3.2, 4]. Microwave technology applications will be considered in Chap. 4.

3.1 Physical Fundamentals

The Cooper pairs of each superconducting electrode of the SIS structure in Fig. 2.1 form a quantum mechanical system i, whose wave function ψ_i is described by (2.8) if both systems are independent of one another. For sufficiently thin layers of insulator, however, the two systems interact with each other. From the originally uncoupled equations (2.8) with $i = 1, 2$, we obtain now the system of weakly coupled differential equations [3.5]

$$\frac{\partial \psi_1}{\partial t} = -\frac{\mathrm{j}}{\hbar}[W_1\psi_1 + K\psi_2] \; ; \quad \frac{\partial \psi_2}{\partial t} = -\frac{\mathrm{j}}{\hbar}[K\psi_1 + W_2\psi_2] \quad . \tag{3.1}$$

The coupling is symmetric and is described by the real constant K. In the present case coupling means that Cooper pairs are exchanged between the superconductors 1 and 2. From Sect. 1.3, using n_c for the density of the Cooper pairs, we can write down the wave functions and their phases Θ_i:

$$\psi_1 = \sqrt{n_{c1}}\mathrm{e}^{\mathrm{j}\Theta_1} \; ; \quad \psi_2 = \sqrt{n_{c2}}\mathrm{e}^{\mathrm{j}\Theta_2} \quad . \tag{3.2}$$

If (3.2) is substituted in (3.1) we obtain

$$\frac{\dot{n}_{c1}}{2\sqrt{n_{c1}}}\mathrm{e}^{\mathrm{j}\Theta_1} + \mathrm{j}\sqrt{n_{c1}}\mathrm{e}^{\mathrm{j}\Theta_1}\dot{\Theta}_1 = -\frac{\mathrm{j}}{\hbar}\left[W_1\sqrt{n_{c1}}\mathrm{e}^{\mathrm{j}\Theta_1} + K\sqrt{n_{c2}}\mathrm{e}^{\mathrm{j}\Theta_2}\right] \tag{3.3}$$

$$\frac{\dot{n}_{c2}}{2\sqrt{n_{c2}}}e^{j\Theta_2} + j\sqrt{n_{c2}}e^{j\Theta_2}\dot{\Theta}_2 = -\frac{j}{\hbar}\left[K\sqrt{n_{c1}}e^{j\Theta_1} + W_2\sqrt{n_{c2}}e^{j\Theta_2}\right] \quad . \tag{3.4}$$

Now let us take the real and imaginary parts, and call the phase difference between the two superconductors $\Theta_2 - \Theta_1 = \varphi$.

$$\frac{1}{2}\frac{\dot{n}_{c1}}{\sqrt{n_{c1}}} = \frac{K}{\hbar}\sqrt{n_{c2}}\sin\varphi \tag{3.5}$$

$$\frac{1}{2}\frac{\dot{n}_{c2}}{\sqrt{n_{c2}}} = -\frac{K}{\hbar}\sqrt{n_{c1}}\sin\varphi \tag{3.6}$$

$$\sqrt{n_{c1}}\dot{\Theta}_1 = -\frac{1}{\hbar}\left[W_1\sqrt{n_{c1}} + K\sqrt{n_{c2}}\cos\varphi\right] \tag{3.7}$$

$$\sqrt{n_{c2}}\dot{\Theta}_2 = -\frac{1}{\hbar}\left[K\sqrt{n_{c1}}\cos\varphi + W_2\sqrt{n_{c2}}\right] \quad . \tag{3.8}$$

From (3.5) and (3.6) we see that $\dot{n}_{c1} = -\dot{n}_{c2}$. This means that Cooper pairs are only exchanged between the superconductors 1 and 2. In the case of $n_{c1} \neq n_{c2}$ the following calculations are rather complicated [3.26]. Therefore in what follows we shall for the sake of simplicity assume equal superconductors with $n_{c1} = n_{c2} =$ const. and assume that there are always enough Cooper pairs to maintain the charge equilibrium. Then $\dot{n}_{c1} = -\dot{n}_{c2}$ describes not the actual rate of change of the Cooper pairs densities, but just a tendency to change.

From (3.5) and (3.6) it now follows that

$$\dot{n}_{c1} = \frac{2K}{\hbar}n_{c1}\sin\varphi = -\dot{n}_{c2} \quad . \tag{3.9}$$

$\dot{n}_{c1}$, multiplied by the charge of a Cooper pair and an effective distance into the electrode, gives the tunnel current density

$$J = J_c \sin\varphi \tag{3.10}$$

with $-1 \leq \sin\varphi \leq 1$ and J_c as the maximum of the density J of the tunnel current, flowing from electrode 2 to electrode 1.

From the difference of (3.7) and (3.8), with $n_{c1} = n_{c2}$, we find that the time derivative of the phase difference φ is

$$\dot{\varphi} = \frac{1}{\hbar}(W_1 - W_2) \quad . \tag{3.11}$$

When $W_1 = W_2$ the phase difference does not vary with time. If, however, there is a voltage difference U between the superconductors 1 and 2, then there is an energy difference $W_1 - W_2 = 2eU$ between the Cooper pairs and

$$\dot{\varphi} = \frac{2eU}{\hbar} \quad . \tag{3.12}$$

Equations (3.10) and (3.12) are called the Josephson equations.

If $U = 0$ the phase φ is constant. The current density J according to (3.10) is a pure direct current density with the maximal value J_c. As we shall see below, J_c can be influenced by an external magnetic field. This is the d.c. Josephson effect.

If there is a d.c. voltage $U = U_0 \neq 0$ on the Josephson junction, then by integration of (3.12)

$$\varphi = \frac{2eU_0}{\hbar}t + \varphi_0 \tag{3.13}$$

with an integration constant φ_0. The current density J then becomes a pure alternating current density

$$J = J_c \sin\left(\frac{2eU_0}{\hbar}t + \varphi_0\right) \tag{3.14}$$

with the Josephson frequency

$$f_J = \frac{2eU_0}{h} = \frac{U_0}{\Phi_0} \quad . \tag{3.15}$$

The frequency of the alternating current is therefore proportional to the d.c. voltage applied. The constant of proportionality is the reciprocal of the flux quantum $\Phi_0 = 2.07\,\mu\text{Vps} = 2.07\,\mu\text{V/GHz}$. This is the a.c. Josephson effect. An aid to remembering (3.15) is the equality of the photon energy hf_J and the Cooper pair energy difference $2eU_0$ between the electrodes.

Microscopic theory [3.6] yields a connection

$$J_c = \text{SCF}\frac{G_n}{A}\frac{\pi\Delta(T)}{2e}\tanh\frac{\Delta(T)}{2k_BT} \tag{3.16}$$

between the maximal or critical Josephson tunnel current density and the tunnelling conductance per unit surface area G_n/A for the quasi-particles at high enough voltages (see Fig. 3.1). In (3.16) SCF is the "strong coupling factor". For materials with weak electron-phonon interaction SCF = 1, for strong interaction SCF < 1. For example, for lead SCF = 0.788 and for tin SCF = 0.911 [3.3].

For small Josephson tunnel junctions the current density J is uniformly distributed over the electrode surface A, and the critical current is given by $I_c = J_cA$. It is the maximal current that can flow through the Josephson junction with $U = 0$. When $U \neq 0$ Josephson currents and quasi-particle currents can flow in an SIS junction. The current-voltage characteristics externally measurable for an SIS structure at $T = 0$ are shown in Fig. 3.1. Beginning with the operating point $U = 0, I = 0$, as the impressed direct current I increases up to the critical current I_c there is at first no voltage between the electrodes. If I_c is surpassed, the voltage jumps to the quasi-particle curve with $U \neq 0$. In order to regain $U = 0$, the current must be reduced to very small values. For $T > 0$ the curve is similar. The knee in the quasi-particle curve is, however, no longer so sharp.

For small areas A and with no external magnetic field φ is independent of position. The position dependency of φ, when an external magnetic field

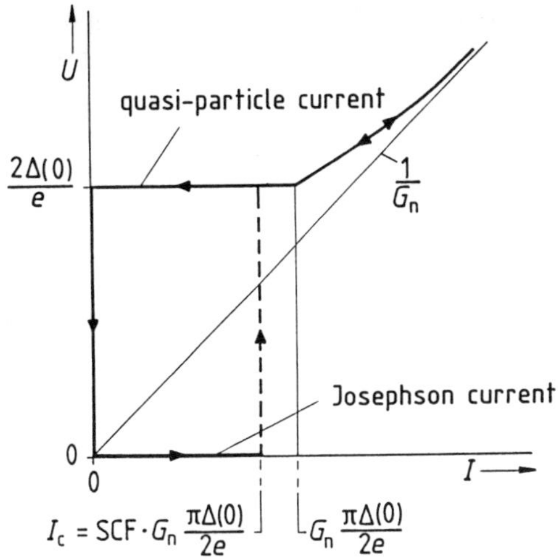

Fig. 3.1. Current-voltage curves of the Josephson junction with the Josephson current and the quasi-particle current for $T = 0$

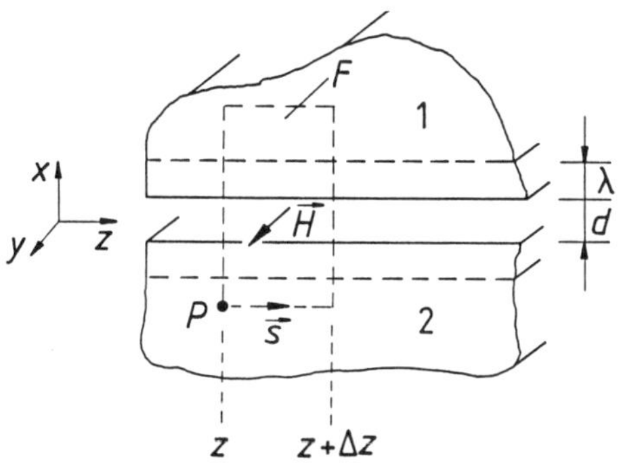

Fig. 3.2. Section through a Josephson tunnel junction illustrating the dependence of the phase φ on the position and the magnetic field

is present, is illustrated in Fig. 3.2. A magnetic field H_y in the y-direction is present between electrodes 1 and 2. Because of the coherent phase coupling in and between the superconductors, the integral over the quantum phase along the closed path $\boldsymbol{s}$ must vanish. If it were not equal to zero, the phase, e.g. at point P, would not be single valued. Using the wave vector $\boldsymbol{k}$ we accordingly have

$$0 = \oint \boldsymbol{k}\, \mathrm{d}\boldsymbol{s} \quad . \tag{3.17}$$

If, as in Sect. 1.5, $\boldsymbol{k}$ is replaced by the canonical momentum from (1.64), we obtain

$$0 = \frac{q_s}{\hbar} \int_{l_1, l_2} \left(\frac{m_s}{n_s q_s^2} \boldsymbol{J}_s + \mu_0 \boldsymbol{A} \right) \cdot \mathrm{d}\boldsymbol{s} + \varphi(z + \Delta z) - \varphi(z) \quad . \tag{3.18}$$

Here $\varphi(z)$ and $\varphi(z + \Delta z)$ are the phase differences between the two superconductors at z and $z + \Delta z$, whilst l_1, l_2 are the integration paths inside the superconductors. l_1 and l_2 should be so deep in the superconductors that the screening currents $\boldsymbol{J}_s$ vanish along the path segments in the directions of z and $-z$. On the path segments in the directions of x and $-x$, $\boldsymbol{J}_s$ is perpendicular to $\boldsymbol{s}$, so that the current contribution to the integral in (3.18) also vanishes. Neglecting the contribution from the tunnelling barrier d and putting $q_s = -2e$ we obtain

$$\Delta\varphi = \varphi(z + \Delta z) - \varphi(z) = \frac{2e\mu_0}{\hbar} \oint \boldsymbol{A} \cdot \mathrm{d}\boldsymbol{s} \quad . \tag{3.19}$$

Using Stokes's law

$$\oint \boldsymbol{A} \cdot \mathrm{d}\boldsymbol{s} = \int\int_F (\nabla \times \boldsymbol{A}) \cdot \mathrm{d}\boldsymbol{F} \quad , \tag{3.20}$$

and (1.63) we then obtain

$$\Delta\varphi = \frac{2e\mu_0}{\hbar} \int\int_F H_y \mathrm{d}F \quad . \tag{3.21}$$

With the effective area $\Delta z(d + 2\lambda)$, and passing from difference to differential quotients, we find that

$$\frac{\partial\varphi}{\partial z} = \frac{2e\mu_0 d'}{\hbar} H_y \tag{3.22}$$

with $d' = d + 2\lambda$. Integration with respect to z gives the tunnel current density from (3.10)

$$J = J_c \sin\left(\frac{2e\mu_0 d'}{\hbar} H_y z + \varphi_0\right) \quad . \tag{3.23}$$

The total current of a Josephson junction bounded by the length l in the z-direction is obtained by a further integration with respect to z. The result is that the maximal current $I_c(H_y)$ that can flow through the Josephson tunnel junction, with $U = 0$, depends on the magnetic field in the following manner:

$$I_c(H_y) = I_c(0) \left| \frac{\sin\pi(\Phi/\Phi_0)}{\pi(\Phi/\Phi_0)} \right| \quad . \tag{3.24}$$

Here Φ_0 is the flux quantum from (1.69) and $\Phi = \mu_0 H_y l d'$ the magnetic flux through the Josephson junction. Figure 3.3 shows the behaviour of (3.24). With a sufficiently strong magnetic field the Josephson current and hence the Josephson effect as a whole can be suppressed.

3.2 Concentrated Josephson Junctions

In this section we shall investigate the d.c. characteristics of concentrated Josephson junctions. We shall take as our basis the RCSJ model (Resistively and Capacitively Shunted Junction), which in most cases describes correctly the relationships measured at the terminals between the current and voltage not only qualitatively, but also quantitatively. According to Fig. 3.4 it consists first of all

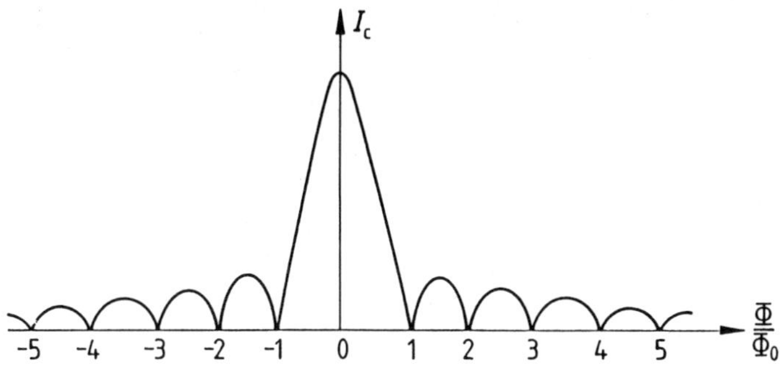

Fig. 3.3. Dependence of the critical current I_c as a function of the magnetic flux Φ flowing through the Josephson tunnel junction. $\Phi_0 = h/2e$ is the flux quantum

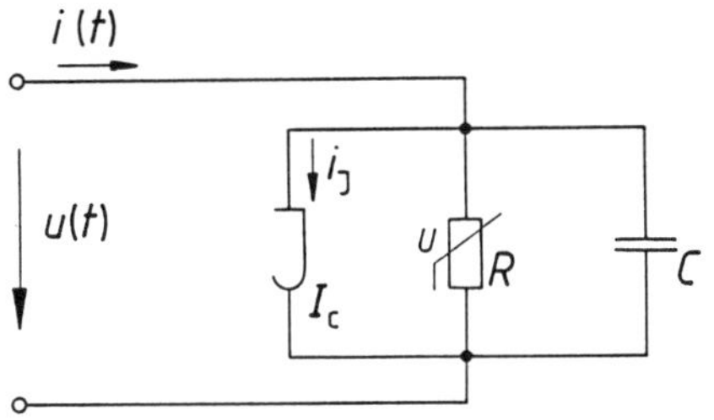

Fig. 3.4. General RCSJ equivalent circuit of a Josephson Junction

of the ideal Josephson junction J, which is uniquely characterised by its critical current I_c, and is described by

$$i_J(t) = I_c \sin(\varphi(t)) \tag{3.25}$$

$$\dot{\varphi}(t) = 2eu(t)/\hbar \quad . \tag{3.26}$$

Parallel to this lies a voltage-dependent nonlinear resistance R which is described, for example, by the quasi-particle tunnelling curve from Fig. 2.3. Finally also in parallel lies the capacitance C which is determined by the arrangement of the electrodes and the dielectric. According to the construction of a Josephson junction, I_c, $R(u)$ and C vary.

For the total current we have

$$i(t) = I_c \sin\varphi + \frac{u}{R} + C\frac{\mathrm{d}u}{\mathrm{d}t} \quad . \tag{3.27}$$

According to the current and voltage applied, $i(t)$ and $u(t)$ in Fig. 3.4 can contain d.c. and a.c. components. First of all we shall study the autonomous case, i.e. only a d.c. component will be assumed to be applied externally.

3.2.1 Autonomous Operation

In the simplest case $C = 0$ and R = const. Thiss case is represented to good approximation by the junctions in Fig. 3.5. In the microbridge, Fig. 3.5a, a superconducting strip is so constricted for a short length that the superconduction there is weakened or completely suppressed. The wave functions of the superconductors in the broad electrodes, however, are coupled with each other via the narrow segment, so that under certain circumstances a Josephson current can flow. In addition, a normally conducting current proportional to the voltage also flows. The capacitance of this arrangement is negligibly small. The electrode leads are often made thicker than the actual bridge region in order to reduce unwanted heat effects (VTB - Variable Thickness Bridge). The three-dimensional analogue of the microbridge, viz. the point contact (see Fig. 3.5b), is effected very simply: one presses a pointed superconducting wire on to a superconducting base-plate. In the region of the contact there are effects similar to those at the microbridge. The critical current I_c at the point contact is adjustable by pressure.

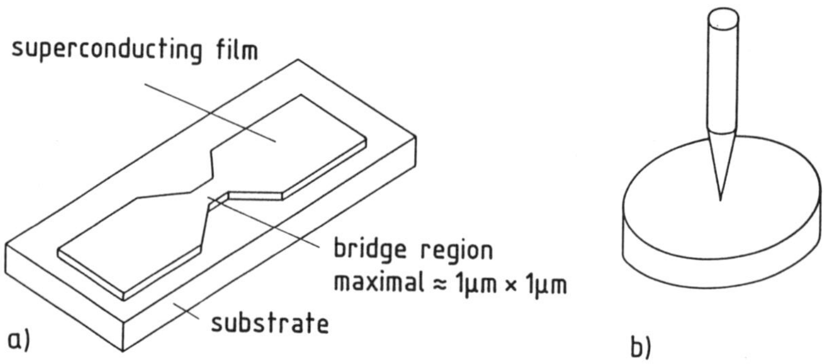

Fig. 3.5a,b. Josephson junctions with $C \simeq 0$ and $R \simeq$ const.; **(a)** microbridge, **(b)** point contact

We shall first of all study the case when only a direct current $i(t) = I$ = const. is passed through the Josephson junction of Fig. 3.4. With $C = 0$ and $R = R_n$ = const. the total current is given by

$$I = i_J + \frac{u}{R_n} = I_c \sin\varphi + \frac{u}{R_n} \quad . \tag{3.28}$$

For $u = 0$ the solution is φ = const. with $-I_c \leq I \leq I_c$. For $u \neq 0$, substituting (3.26) in (3.28), putting $\dot{\varphi} = \mathrm{d}\varphi/\mathrm{d}t$ and separating the variables, we get

$$\frac{\mathrm{d}\varphi}{I/I_c - \sin\varphi} = \frac{2eR_nI_c}{\hbar}\mathrm{d}t \quad . \tag{3.29}$$

Now when both sides are integrated one obtains $\varphi(t)$. From (3.26) $u(t)$ is then found to be

$$u(t) = I_cR_n \frac{(I/I_c)^2 - 1}{I/I_c + \sin\left[\sqrt{(I/I_c)^2 - 1}\cdot R_nI_c2et/\hbar + \varphi_0\right]} \tag{3.30}$$

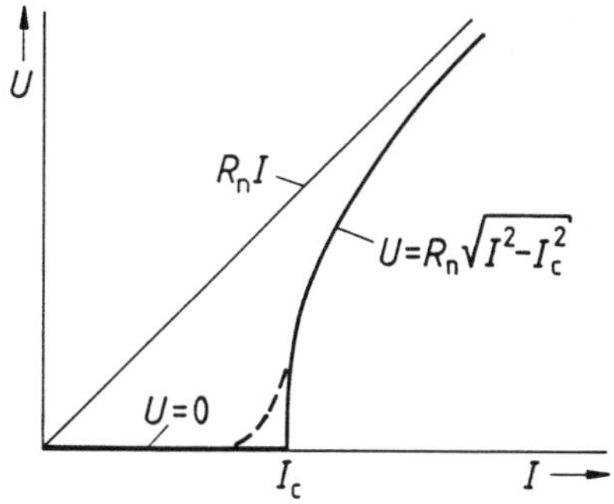

Fig. 3.6. Current-voltage curve for $R = R_n$ = const., $C = 0$ and for impressed direct current. (- - -) rounding off by noise at $\hbar I_c/ek_BT = 40$

with the constant φ_0. The d.c. voltage U at the junction is equal to the arithmetic mean value of (3.30). It is found to be

$$U = R_n\sqrt{I^2 - I_c^2}, \quad |I| > I_c \quad . \tag{3.31}$$

Figure 3.6 shows the resulting curve. If $u(t) = 0$ then also the d.c. voltage $U = 0$ up to the critical current I_c. Thereafter it increases like a square root and approaches asymptotically the straight line R_nI. Figure 3.7 shows $u(t)$ according to (3.30) at two different operating points. If $I \gg I_c$ the a.c. voltage contribution in u is almost sinusoidal, whereas it is greatly distorted if I is only somewhat larger than I_c. From (3.30) it can be seen that in each case the period is the reciprocal of the Josephson frequency, the d.c. voltage being taken from (3.31).

In Josephson junctions with $C \neq 0$, according to (3.27) and Fig. 3.4 and in contrast to (3.28), the total current includes a displacement current. No analytic expression is known for the direct current characteristic with an impressed direct current $i(t) = I$ = const. Numerically calculated results for $R = R_n$ = const. are displayed in Fig. 3.8. The size of the capacitance is here expressed by the McCumber parameter β_c [3.8]:

$$\beta_c = \omega_c C R_n \quad \text{with} \quad \omega_c = 2eI_cR_n/\hbar \quad . \tag{3.32}$$

Here the characteristic frequency $\omega_c/2\pi$ is the Josephson frequency at the voltage given by the product of the critical current I_c and the normal resistance R_n. β_c is thus the product of the time constant R_nC and the angular frequency ω_c.

Figure 3.8a shows the direct current curves for three values of β_c. In contrast to Fig. 3.6 a cycle with increasing and decreasing currents can cause hysteresis to occur. According to Fig. 3.8b this hysteresis vanishes for sufficiently small capacitance, i.e. roughly for $\beta_c \leq 1$. When $\beta_c \to \infty$ the alternating current arising from the sin φ term is short-circuited by the capacitance. The a.c. voltage is zero. Then the current from the sin φ term is purely sinusoidal and does not contribute to the direct current (see Fig. 3.8a).

Josephson junctions with SIS structure as in Fig. 2.1 have a nonvanishing capacitance $C \neq 0$ and a voltage-dependent nonlinear resistance $R(u)$. Under these conditions the calculation of the RCSJ model becomes still more complicated. An example of a numerical solution [3.9] is shown in Fig. 3.9. The quasi-particle

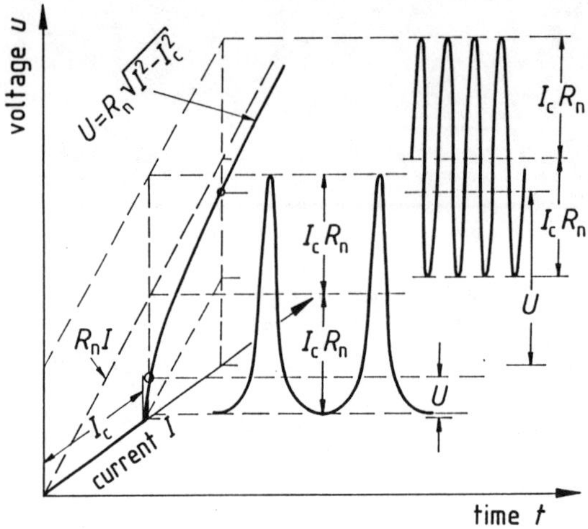

Fig. 3.7. Time-dependent voltage $u(t)$ at two operating points on the curve of Fig. 3.6, according to [3.7]

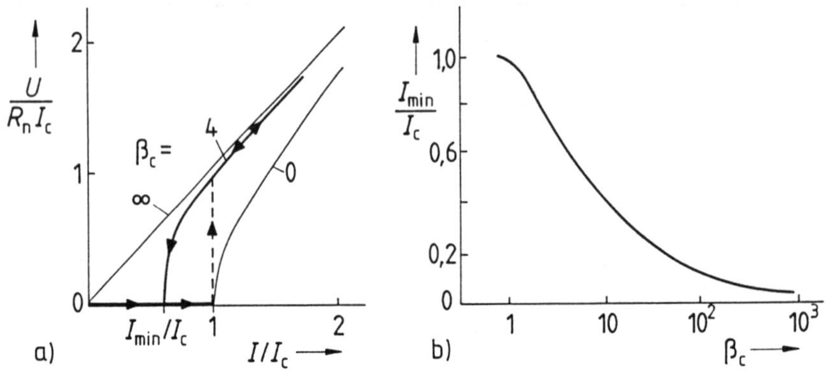

Fig. 3.8a,b. Direct current curves for $R = R_n$ = const., $C \neq 0$ and for impressed direct current according to [3.8]. (**a**) Curve with $\beta_c = 4$. Limiting cases $\beta_c = 0$ and ∞ also indicated. (**b**) Current ratio I_{min}/I_c as a function of the McCumber parameter β_c

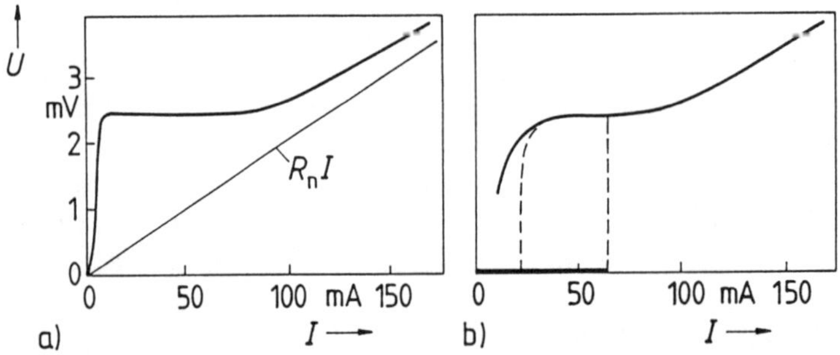

Fig. 3.9a,b. Direct current curves of Pb/insulator/Pb tunnel junction at 4.2 K according to [3.9]. (**a**) Measured quasi-particle tunnel current ; Josephson effect suppressed. (**b**) (—) calculated and (- - -) measured curves; Josephson effect not suppressed

characteristic, Fig. 3.9a, the calculated and the measured profiles of which are practically identical, provided the functional dependence of the resistance on the d.c. voltage $R(U)$. Using the measured I_c and the estimated C, the direct current characteristic was then calculated. The continuous curve in Fig. 3.9b shows the result. It agrees with the experimental result, except that the sudden return to voltage $U = 0$ occurs at somewhat greater currents. Noise effects are presumably responsible for this. We see that the knee at the energy gap voltage is distinctly rounded. This is explained by the fact that the alternating current arising from the $\sin \varphi$ term first of all causes an a.c. voltage drop at C, and this is rectified by the nonlinear resistance $R(U)$ when the d.c. voltage is near to the gap voltage. This causes an additional direct current [3.10].

Noise effects in Josephson junctions can often be treated as noise in the parallel resistance $R(U)$. This is understandable, because the $\sin \varphi$ term in Fig. 3.4 is loss-free. Depending on the nature of the parallel resistance, one has to take account of the noise either for the tunnel junction according to (2.14), or as thermal noise in the ohmic parallel resistance at point contacts and microbridges. The noise causes the knee in the characteristic curve to be rounded. In the RSJ model with $C = 0$ the parameter $\gamma = \hbar I_c / e k_B T$ is the critical quantity. The rounding is shown in Fig. 3.6 for the case when $\gamma = 40$ [3.11].

A further source of noise can arise from the interaction of the pair current in the $\sin \varphi$ term with its environment. This is then described by a noise power density according to (2.14), where e is replaced by $2e$. It is, however, not yet quite clear under what circumstances this source of noise has to be taken into account.

3.2.2 Microwave Injection

As we have already established above, an applied d.c. voltage causes an a.c. current to flow through a Josephson junction. It therefore behaves as an oscillator. An externally applied oscillation can cause an oscillator to be synchronised to the basic oscillation and to the harmonics of the external source. In a Josephson junction such synchronised states are recognisable as steps in the direct current characteristic, called Shapiro steps after their discoverer. We shall now explain this phenomenon.

We shall again take the RCSJ model of Fig. 3.4 as our basis and shall assume that the junction has externally applied to it a superposition of a d.c. voltage U_0 with only one sinusoidal a.c. voltage:

$$u(t) = U_0 + \hat{U}_1 \cos \omega t \quad . \tag{3.33}$$

Integration of (3.26) and substitution in (3.25) yields the current by the $\sin \varphi$ term:

$$i_J(t) = I_c \sin \left\{ \frac{2eU_0}{\hbar} t + \frac{2e\hat{U}_1}{\hbar\omega} \sin(\omega t) + \varphi_0 \right\} \tag{3.34}$$

with an integration constant φ_0. Just as for frequency modulation [3.12], (3.34) can be expanded in a Fourier-Bessel series:

$$i_J(t) = I_c \sum_{m=-\infty}^{\infty} J_m\left(\frac{2e\hat{U}_1}{\hbar\omega}\right) \sin\left\{\left(m\omega + \frac{2eU_0}{\hbar}\right)t + \varphi_0\right\} \tag{3.35}$$

with $J_m(x)$ the Bessel function of order m.

Then $i_J(t)$ has a non-zero d.c. component I_J only if $2eU_0/\hbar = n\omega$ for some integer n. We then obtain

$$I_J = (-1)^n I_c J_n\left(\frac{2e\hat{U}_1}{\hbar\omega}\right) \sin\varphi_0 = (-1)^n I_c J_n\left(n\frac{\hat{U}_1}{U_0}\right) \sin\varphi_0 \quad . \tag{3.36}$$

This equation is formally very similar to the Josephson equation (3.25). In both equations the direct current is determined by quantum-mechanical phase differences. From (3.35) we see that φ_0 is the phase difference between two oscillators with equal frequencies. The one is the mth harmonic of the externally applied oscillation. The other is the Josephson oscillation. If both oscillators are locked in phase, there is a direct current whose strength can lie between $\pm\Delta I_n = \pm I_c|J_n(2e\hat{U}_1/\hbar\omega)|$. The current curve therefore forms steps. These are shown in Fig. 3.10a for $2e\hat{U}_1/\hbar\omega = 4.7$.

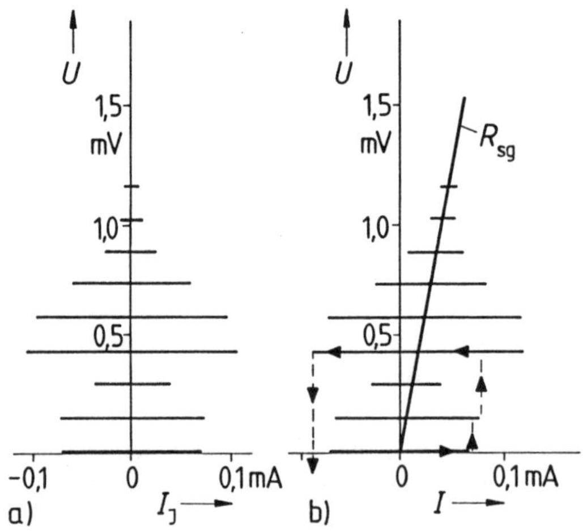

Fig. 3.10a,b. Microwave-induced steps with an applied voltage. f = 70 GHz; $\hat{U}_1$ = 0.68 mV ($2e\hat{U}_1/\hbar\omega$ = 4.7); I_c = 0.25 mA; R_{sg} = 26 Ω, from [3.10], **(a)** direct current I_J through the sin φ term, **(b)** total direct current I through the Josephson junction

Figure 3.10b shows the addition of the direct current through the resistance in parallel $R(U)$ of Fig. 3.4. $R(U)$ is here replaced by a relatively high linear resistance R_{sg} which for a Josephson tunnel junction will reproduce the small quasi-particle current at voltages below the energy gap voltage. The direct current characteristic consists of many branches. For an impressed direct current it is many-valued. The choise of an operating point then depends very much on the previous history. The variation of the operating point with a special current curve is illustrated in Fig. 3.10b. Beginning with $U = 0, I = 0$, the current I was first increased to about 70 μA and then decreased. Steps which intersect the voltage axis are called zero current steps. At operating points which lie to the left of the centres of the steps, RF power of the external generator is transformed into d.c. power in the sin φ term.

The assumption that the a.c. voltage across the Josephson junction is purely sinusoidal is correct for large capacitance C, as may happen in Josephson tunnel junctions with SIS structure. All the harmonics arising in the sin φ term are then short-circuited. Then even the size of the internal resistance of the RF generator has no effect on the shape of the direct current characteristic. With capacitance of finite size this is in general no longer true.

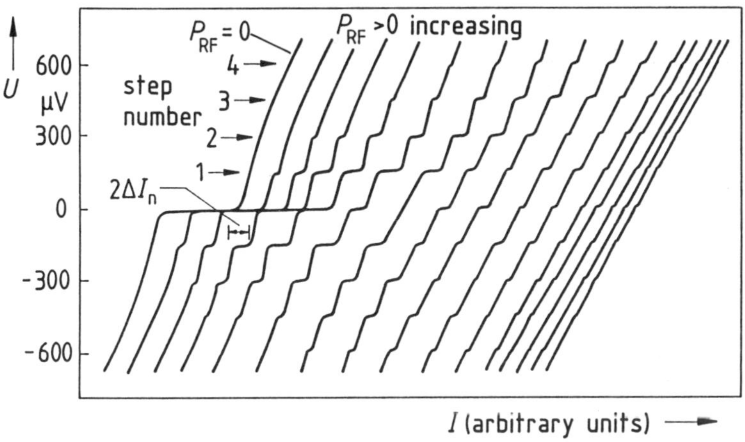

Fig. 3.11. Microwave-induced steps at an Nb point contact for various microwave powers P_{RF}, f = 72 GHz, T = 4.2 K, from [3.13]

In the limiting case with $C = 0$ and $R(U) = R_n$ =const. there is no longer in practice any hysteresis. Figure 3.11 shows a typical example [3.13]. The steps of constant voltage in each of the curves are displaced so far from each other in the direction of the current that they no longer overlap. The step width $2\Delta I_n$ in general no longer depends on the critical current and on the RF amplitude as predicted by (3.36), but behaves as shown in Fig. 3.12. This compares the measured values [3.14] with the calculated values [3.15], based on an impressed current.

$$i(t) = I_0 + \hat{I}_1 \cos \omega t \quad , \tag{3.37}$$

together with $C = 0$ and $R(U) = R_n$ = const. The frequency ratio ω/ω_c, with ω_c from (3.32), appears as a parameter. For small values of ω/ω_c, corresponding to large values of R_n, the step widths $2\Delta I_n$ are far less pronounced than for $\omega/\omega_c = 1$. The latter corresponds to small values of R_n, so that there is almost an impressed RF voltage across the sin φ term, and one again achieves a step width predicted by the Bessel functions in (3.36).

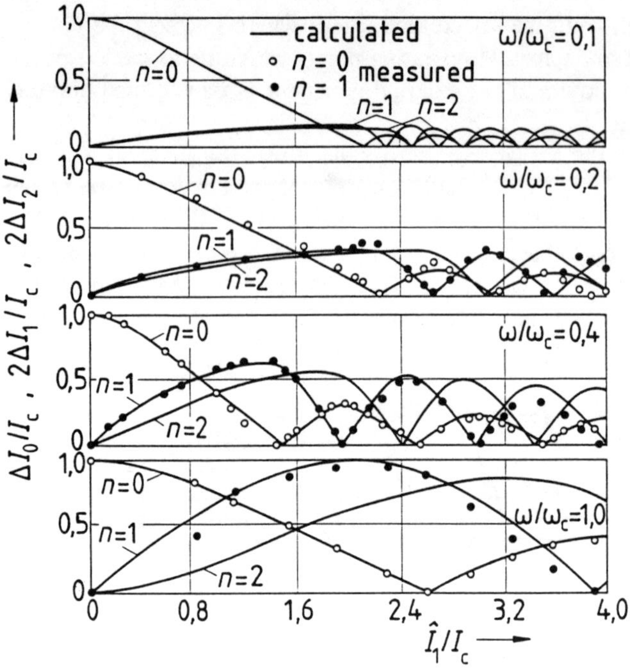

Fig. 3.12. Calculated [3.15] and measured [3.14] dependence of the step width ΔI_n of a Josephson junction with applied RF current. Measurements at 1 GHz on molybdenum microbridges: the maximal values of $2J_n(x)$ are 1.16 for $n = 1$ and 0.972 for $n = 2$

With a non-zero parallel capacitance $C \neq 0$, and in certain regions of values of the applied RF power and its frequency, the direct current curves of Josephson junctions reveal some irregular zones, such as, for example, intermittent steps and regions of negative differential resistance. They are explained by the fact that the differential equation (3.27) then has chaotic solutions.

In [3.16, 17] conditions are given for freedom from chaos in Josephson junctions. Among other things, it is required that for the frequency f of the injected microwave signal $f/f_\mathrm{p} \gg 1$, where

$$f_\mathrm{p} = \frac{1}{2\pi}\sqrt{\frac{2eI_\mathrm{c}}{\hbar C}} = \frac{1}{2\pi}\sqrt{\frac{2eJ_\mathrm{c}d}{\hbar\varepsilon_0\varepsilon_\mathrm{r}}} \tag{3.38}$$

is called the plasma frequency of the Josephson junction (see also Sect. 3.3).

The intrinsic noise of the Josephson junction and also external noise cause the edges of the microwave-induced steps to be rounded [3.18]. The centres of the steps, however, lie at voltages which are integral multiples of $hf/2e$.

3.3 Distributed Josephson Tunnel Junctions

With Josephson tunnel junctions which are not short in their lateral dimensions, additional peculiarities occur. In this section we shall summarise these peculiarities. We shall consider a tunnel junction as in Fig. 3.2, and for the sake of simplicity shall assume that it is extensive only in the z-direction, but is narrow in the y-direction. The most important effects then already become apparent.

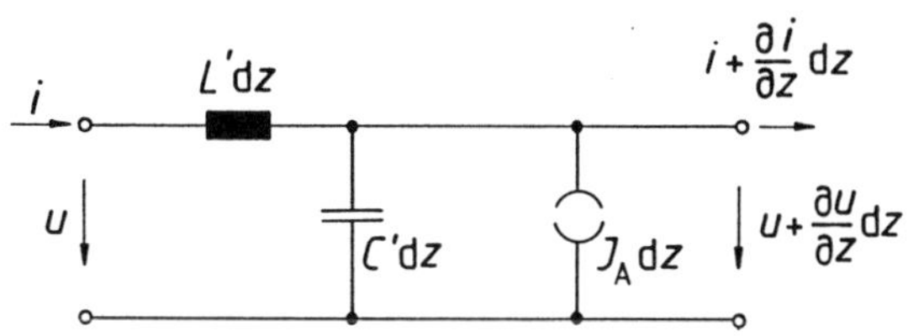

Fig. 3.13. Equivalent circuit for loss-free distributed Josephson tunnel junction

We can regard the Josephson tunnel junction as a transmission line in the z-direction. The transmission line element of length Δz is then represented by the equivalent circuit of Fig. 3.13 with the capacitance per unit length C' and the inductance per unit length L'. With the y-dimension w and the relative permittivity ε_r of the insulating layer we then have

$$C' = \frac{\varepsilon_0 \varepsilon_r w}{d} \; ; \; L' = \frac{\mu_0 d'}{w} \quad , \tag{3.39}$$

where the inductance per unit length is not determined by the plate separation d alone, but by the effective separation $d' = d + 2\lambda$, because the magnetic field penetrates into the superconductor. The surface current density J_A in Fig. 3.13 is given by

$$J_A(z,t) = w J_c \sin \varphi(z,t) \quad . \tag{3.40}$$

Starting from this equivalent circuit we can in the usual way [3.19] set up the transmission-line differential equations and hence obtain the wave equation of the transmission line. With

$$v = \frac{1}{\sqrt{L'C'}} = \frac{\sqrt{d/d'}}{\sqrt{\mu_0 \varepsilon_0 \varepsilon_r}} \tag{3.41}$$

we obtain the wave equation for the voltage

$$\left(\frac{\partial^2}{\partial z^2} - \frac{1}{v^2} \frac{\partial^2}{\partial t^2} \right) u = L' \frac{\partial J_A}{\partial t} \quad . \tag{3.42}$$

If this equation is integrated with respect to time and account is taken of (3.26), (3.29) and (3.40), we obtain the wave equation for the phase difference

$$\left(\frac{\partial^2}{\partial z^2} - \frac{1}{v^2}\frac{\partial^2}{\partial t^2}\right)\varphi = \frac{1}{\lambda_J^2}\sin\varphi \quad , \tag{3.43}$$

where

$$\lambda_J = \sqrt{\frac{\hbar}{2e\mu_0 d' J_c}} \tag{3.44}$$

is called the Josephson penetration depth, since in the zero-voltage state $u = 0$, corresponding to $\partial\varphi/\partial t = 0$, and for small phase differences $\varphi \simeq \sin\varphi$, (3.43) becomes

$$\frac{\partial^2\varphi}{\partial z^2} = \frac{1}{\lambda_J^2}\varphi \quad , \tag{3.45}$$

as in (1.20). The solutions $\varphi \sim \exp(-z/\lambda_J)$ and $\varphi \sim \exp(z/\lambda_J)$ describe the current distributions $J_A(z)$ in the Josephson tunnel junction which die away exponentially towards the middle from the ends at $z = 0$ and $z = l$ with penetration depth λ_J. A typical value for λ_J, for example, with $J_c = 50\,\mathrm{A/cm^2}$ and $d' = d + 2\lambda \simeq 2\lambda = 80\,\mathrm{nm}$, is given by $\lambda_J = 81\,\mu\mathrm{m}$.

For critical current densities J_c which are small compared with the displacement current density, i.e. for

$$|J_c| \ll \frac{C'}{w}\left|\frac{\partial u}{\partial t}\right|_{\max} \tag{3.46}$$

the right-hand side in (3.42) can be set approximately equal to zero. We then obtain

$$\left(\frac{\partial^2}{\partial z^2} - \frac{1}{v^2}\frac{\partial^2}{\partial t^2}\right)u = 0 \quad , \tag{3.47}$$

and hence the general wave equation. Its solutions are travelling waves of arbitrary form, propagating with the velocity v [3.19]. With pure harmonic time dependence the waves have the phase constant $\beta = \omega/v$ (see Fig. 3.14).

Another limiting case occurs if the phase difference φ depends only weakly on time, i.e. if the electromagnetic wave in the Josephson tunnel junction is small in amplitude. Then φ can be separated into a space-dependent and a small space- and time-dependent component:

$$\varphi(z,t) = \varphi_0(z) + \varphi_1(z,t) \quad . \tag{3.48}$$

From (3.43), with $\cos\varphi_1 \simeq 1$ and with $\sin\varphi_1 \simeq \varphi_1$ we thus obtain for the time-dependent portion

$$\left(\frac{\partial^2}{\partial z^2} - \frac{1}{v^2}\frac{\partial^2}{\partial t^2}\right)\varphi_1 = \frac{1}{\lambda_J^2}\varphi_1\cos\varphi_0 \quad , \tag{3.49}$$

and for pure harmonic time dependency with $\partial^2/\partial t^2 = -\omega^2$

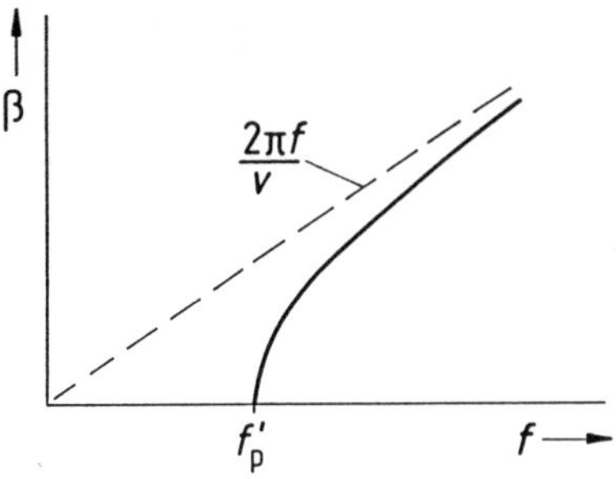

Fig. 3.14. Dispersion characteristic of a Josephson tunnel junction, (f) frequency, (β) phase constant

$$\left[\frac{\partial^2}{\partial z^2} + \left(\frac{\omega^2}{v^2} - \frac{\cos\varphi_0}{\lambda_J^2}\right)\right]\varphi_1 = 0 \quad . \tag{3.50}$$

Equation (3.50) describes a wave propagation with the phase constant

$$\beta = \sqrt{\frac{\omega^2}{v^2} - \frac{\cos\varphi_0}{\lambda_J^2}} = \frac{\omega}{v}\sqrt{1 - \left(\frac{\omega_p'}{\omega}\right)^2} \quad . \tag{3.51}$$

Hence, abbreviating,

$$\omega_p' = \frac{v}{\lambda_J}\sqrt{\cos\varphi_0} = \sqrt{\frac{2eJ_c d}{\hbar\varepsilon_0\varepsilon_r}}\sqrt{\cos\varphi_0} = \omega_p\sqrt{\cos\varphi_0} \tag{3.52}$$

with $\omega_p/2\pi = f_p$ from (3.38). For $\omega < \omega_p'$, β in (3.51) is imaginary and describes an aperiodic attenuation. The wave becomes capable of propagating only for real β, i.e. for $\omega > \omega_p'$ (see Fig. 3.14). There is a formally identical dispersion relationship in the metal waveguide and in the plasma. Accordingly $f_p' = \omega_p'/2\pi$ is called the plasma frequency. Occasionally with the Josephson tunnel junction one understands by this plasma frequency f_p according to (3.37), where in (3.52) $\sqrt{\cos\varphi_0}$ is tacitly taken equal to 1.

Equation (3.43) is the sine-Gordon equation. It crops up in several regions of physics and describes wave propagation in a nonlinear dispersive medium. Wave packets which do not lose their shape even with propagation losses are solutions of the sine-Gordon equation. They are known as solitons [3.20]. Such solitons propagate in distributed Josephson tunnel junctions [3.3] and can also form standing waves. The occurrence of solitons can also influence the direct current characteristic of a Josephson tunnel junction. These effects, however, will not here be elaborated on further.

In the transmission line equivalent circuit Fig. 3.13 losses can be considered by an additional series resistance per unit length and a shunt conduction per unit length, caused by the finite RF conductivity of the superconducting electrodes, as well as by tunnelling of individual electrons and by ohmic currents between the electrodes [3.21]. Then further terms arise in (3.42) and (3.43).

The functioning of distributed Josephson tunnel junctions is strongly influenced by applied external magnetic fields. In calculations these magnetic fields can in principle be allowed for as in (3.22) and (3.23). In particular, with a d.c.

voltage U on the junction and a d.c. magnetic field $H_y = -H_0$, we find from (3.12) and (3.22) that

$$\varphi(z,t) = \omega_J t - k_z z \tag{3.53}$$

with

$$\frac{\partial \varphi}{\partial t} = \omega_J = \frac{2eU}{\hbar} \tag{3.54}$$

and

$$-\frac{\partial \varphi}{\partial z} = k_z = \frac{2ed'\mu_0 H_0}{\hbar} \quad . \tag{3.55}$$

Equation (3.53) describes a phase wave with angular frequency ω_J travelling in the z-direction with the wave number k_z.

3.4 Superconducting Loops with Josephson Junctions

A small superconducting loop containing one or several Josephson junctions is known as a SQUID. This acronym stands for "**S**uperconducting **QU**antum **I**nterference **D**evice". Using SQUIDs the smallest changes in magnetic fields can be measured more sensitively than with any other device. Variants are also suitable for the measurement of magnetic field gradients, voltages, currents and magnetic susceptibilities. In this section we shall discuss the basic method of operation of the SQUID.

Figure 3.15 shows a superconducting loop containing two Josephson junctions. We wish to find the total current I_g that can flow through the loop without a d.c. voltage drop, and how this maximal current depends on the magnetic flux Φ_F passing through the loop. Just as in Sect. 1.5 we shall start out from the proposition that in the zero-voltage state the wave function (1.33) must be uniquely determined at every point of the loop to within a phase factor exp (j$2n\pi$). For the integral of the wave number $\boldsymbol{k}$ of the Cooper pairs along the closed path $\boldsymbol{s}$ we must accordingly have

$$2n\pi = \oint \boldsymbol{k} \cdot \mathrm{d}\boldsymbol{s} \quad . \tag{3.56}$$

If we denote the phase differences in the Josephson junctions 1 and 2 by φ_1 and φ_2 and replace the wave number $\boldsymbol{k}$ in the superconductors by the momentum $\boldsymbol{p}_{co}$ of the Cooper pairs according to (1.3), we get

$$2n\pi = \varphi_2 - \varphi_1 + \frac{1}{\hbar}\int_{u,l} \boldsymbol{p}_{co} \cdot \mathrm{d}\boldsymbol{s} \quad . \tag{3.57}$$

Here we integrate over the upper (u) and the lower (l) halves of the loop. In order to express the integral in (3.57) by the magnetic flux Φ_F, we shall make use

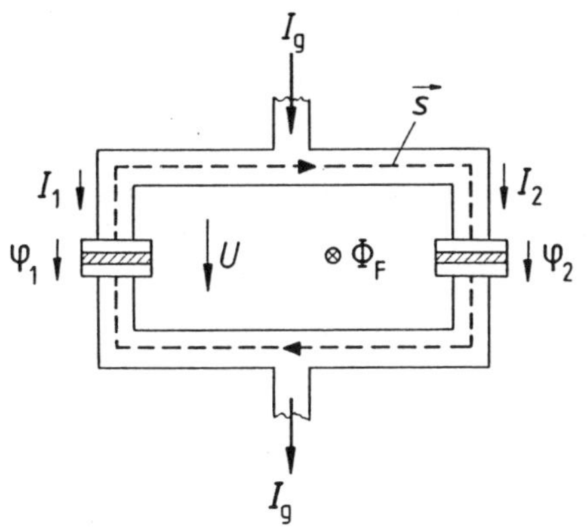

Fig. 3.15. Josephson junctions connected in parallel in a closed loop. The current supply is arranged symmetrically. (s integration path

of the result of our calculation between (1.62) and (1.68). If we choose the path $\boldsymbol{s}$ so deep in the interior of the superconductor that the superconducting current vanishes, then we get the result

$$\int_{u,l} \boldsymbol{p}_{co} \mathrm{d}\boldsymbol{s} = 2e\Phi_F \quad . \tag{3.58}$$

In what follows we shall no longer take account of the term $2n\pi$ in (3.57), since the current through the Josephson junction 2, namely $I_{c2} \sin \varphi_2$, does not change if we add integral multiples of 2π to φ_2. Then we obtain from (3.57) and (3.58)

$$\varphi_2 = \varphi_1 - \frac{2\pi\Phi_F}{\Phi_0} \quad , \tag{3.59}$$

where $\Phi_0 = h/2e$ represents the flux quantum. The total current I_g through both Josephson junctions connected in parallel is accordingly

$$\begin{aligned} I_g &= I_1 + I_2 = I_{c1} \sin \varphi_1 + I_{c2} \sin \varphi_2 \\ &= I_{c1} \sin \varphi_1 + I_{c2} \sin (\varphi_1 - 2\pi\Phi_F/\Phi_0) \quad . \end{aligned} \tag{3.60}$$

The maximal direct current I_{gc} without voltage drop is found by determining the maximum of (3.60) with respect to variation of φ_1, by differentiating and equating to zero. For simplification we shall regard the flux Φ_F as an externally impressed quantity, and so we shall ignore the flux induced by the loop current itself. For a symmetric arrangement with $I_{c1} = I_{c2}$ this maximisation yields

$$I_{gc}(\Phi_F) = 2I_{c1} \left| \cos \frac{\pi\Phi_F}{\Phi_0} \right| \tag{3.61}$$

which is depicted in Fig. 3.16. The critical current I_{gc} of our pair of Josephson junctions can thus be varied by 100% between $2I_{c1}$ and 0 by means of the magnetic field. The current-voltage characteristic of the parallel circuit at these extreme values is illustrated in Fig. 3.17.

In practical SQUIDs the self-induced flux from the loop current is usually not negligible. This means that I_{gc} can then no longer be modulated through 100%

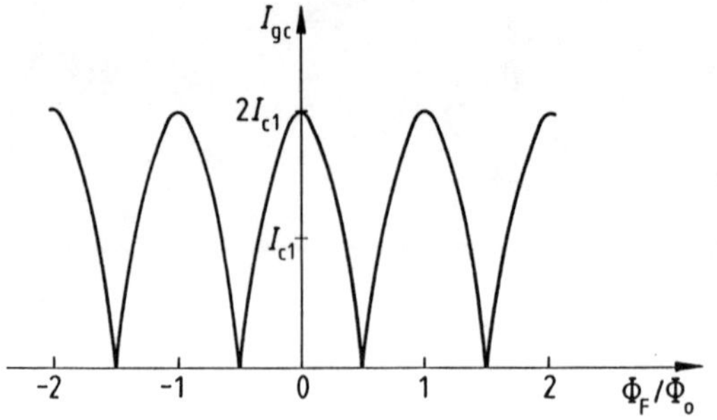

Fig. 3.16. Functional dependence of the critical current I_{gc} of the arrangement of Fig. 3.15 on the applied flux Φ_F. It is assumed that $I_{c1} = I_{c2}$; self-induced flux is neglected

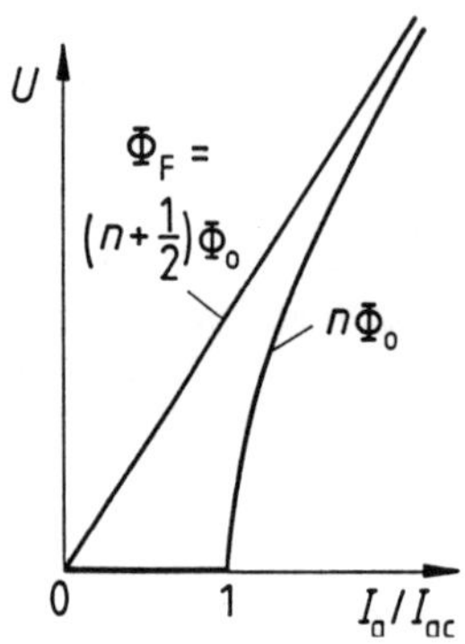

Fig. 3.17. Current-voltage characteristic for the arrangement of Fig. 3.15 with $I_{c1} = I_{c2}$ and neglect of the self-induced flux. Parameter: impressed magnetic flux Φ_F

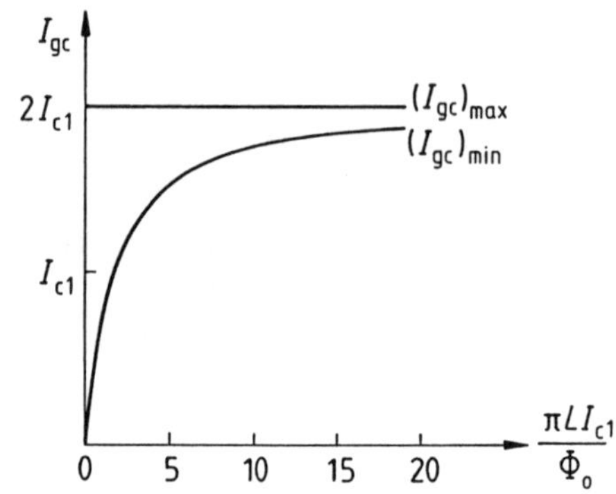

Fig. 3.18. Maximal and minimal values of the critical current I_{gc} as a function of the normalised loop inductance L. The difference between these curves indicates the depth of modulation of I_{gc} attainable by varying the applied magnet field, from [3.2]

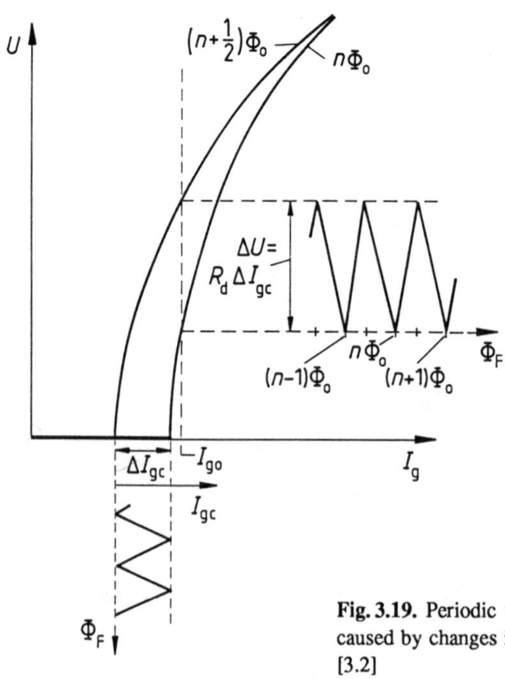

Fig. 3.19. Periodic voltage variations across the DC-SQUID, caused by changes in the applied magnetic field, according to [3.2]

by the externally applied flux. Figure 3.18 shows how the maximal and minimal values of the critical current I_{gc} depend on the inductance L of the loop. The larger L is, the smaller is the depth of modulation. Even with an asymmetric arrangement, i.e. $I_{c1} \neq I_{c2}$, a 100 % modulation is unattainable.

A magnetometer formed by the parallel circuit of two Josephson junctions as in Fig. 3.15 is called the direct current or DC-SQUID. Its current-voltage characteristic taking into account the self-induced flux is shown in Fig. 3.19. With an applied current I_{g0}, variation of the external magnetic field modulates the voltage drop at its peak by about $\Delta U = R_d \Delta I_{gc}$. Here R_d is the mean differential resistance and ΔI_{gc} is the modulation depth of the critical current. In practical SQUID measuring instruments, by means of a negative feed-back with an additional coil, the flux through the SQUID loop, and hence also I_{gc} and ΔU, is held constant. The strength of the feed-back current necessary for this is then a measure of the magnetic field to be measured. A particularly high sensitivity is achieved if a low frequency signal is superposed on the feed-back current, and the low frequency voltage thereby caused is detected by a lock-in amplifier. With the help of superconducting flux transformers one can also measure magnetic fields which do not lie at the location of the SQUID loop. With an optimally designed SQUID such high sensitivities are possible that they correspond to an energy resolution of the magnetic field of about 1×10^{-33} Joules per Hertz of measured band width. That is about twice Planck's constant h.

Arrangements with two or three Josephson junctions connected in parallel, also known as interferometers, are interesting building blocks for digital circuits. The switching over between different states is possible in extremely short times with very small energies.

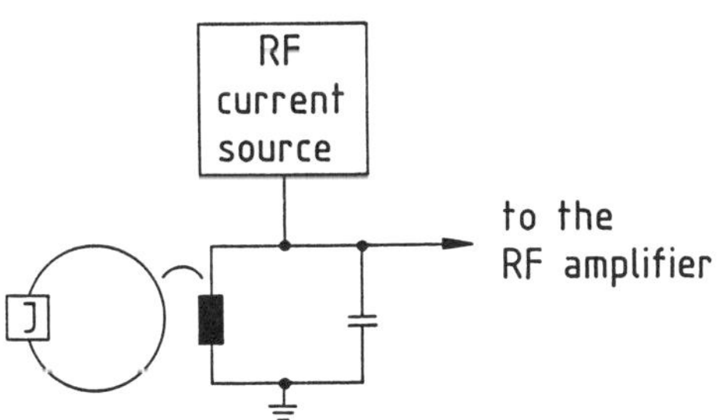

Fig. 3.20. Circuit diagram of the principle of a radio frequency SQUID

Superconducting magnetometers are made not only from DC-SQUIDs but also from radio frequency or RF-SQUIDs. The principles of their construction are shown in Fig. 3.20. The superconducting loop contains only one Josephson junction. It is magnetically coupled with a parallel resonance circuit, through which flows an impressed radio frequency current, for example, with a frequency of 30 MHz. If now a magnetic d.c. field, threading the loop, is varied in strength, then the loading on the resonance circuit varies. This variation changes periodically with the applied magnetic flux through the loop. This produces also a periodic variation in the radio frequency voltage across the resonance circuit. This voltage is fed to an amplifier and rectified. As in measuring systems with DC-SQUIDs, here also the total flux through the loop is held constant by a negative feed-back. The feed-back current necessary for this is again a measure of the magnetic field variations to be measured. The sensitivity using an RF-SQUID is also increased if the feed-back current has superposed on it a low frequency signal and the detection is carried out by means of a lock-in amplifier.

For the further study of SQUIDs and their applications, [3.2, 3.3, 3.22 and 3.23, p. 729–89] should also be consulted. In particular, the use of SQUIDs for the measurement of biomagnetic signals is described in [3.24] and [3.23, pp. 891–962]. Fundamental hints on the use of Josephson junctions and SQUIDs in digital circuits are to be found in [3.2], and a few particular problems are described in [3.25].

4. Applications of Josephson Junctions in Microwave Engineering

In this chapter a few selected application possibilities of Josephson junctions in microwave engineering will be described. Attention will be focussed on Josephson voltage standards, since they have already found a secure place in the technique, namely in precision measurement technique. Josephson junctions as components in microwave receiver devices indeed reveal very interesting possibilities and have often already been successfully tested in the laboratory, but they are not yet in regular use. They are accordingly treated here in somewhat less detail.

4.1 Josephson Voltage Standards

Voltage standards which work on the basis of the a.c. Josephson effect are routinely used universally in national institutes of metrology as primary standards [4.1]. In these voltage standards one makes use of the frequency-voltage conversion in the Josephson junction, according to (3.15). The constant of proportionality between the accurately determined frequency f and the d.c. voltage U_0, as the target quantity, is equal to the quotient of fundamental physical constants $h/2e$. For metrological purposes this has the internationally agreed value [4.2]

$$K_{J-90} = 483597.9 \frac{\text{GHz}}{\text{V}} \quad . \tag{4.1}$$

The standard voltage experienced at an individual Josephson junction has a maximum in the order of magnitude of the energy gap voltage, i.e. a few millivolts. In order to obtain output voltages in the region of 1 V, one connects in series many d.c. Josephson tunnel junctions operating in the region of the zero-current steps. The monolithically integrated superconducting millimetre wave circuits used for this will first of all be generally described here with regard to their design criteria. We shall first discuss the individual tunnel junction, then the complete circuit on the chip and finally a few measured results.

According to Sect. 3.2.2. microwave injection gives rise in the Josephson junction to steps of constant voltage U_n, with

$$U_n = n \frac{h}{2e} f \quad . \tag{4.2}$$

In order not to have to set the operating point of each individual junction separately in a series of circuits, it is convenient to choose the operating conditions so that the steps for all the junctions, plotted against the direct current I, considerably overlap. This is the case in the region of zero-current steps (see Fig. 3.10). The step widths $2\Delta I_n$ should in addition be as large as possible, so that the intrinsic noise of the Josephson junctions and external disturbing noise currents do not impair the operation of the voltage standard.

In what follows we shall denote by n that step of constant voltage at a Josephson junction for which the voltage standard is designed. In practical operation this particular step will not be set for all the junctions. A few will have further integral step numbers somewhat above or below n. The total voltage is however then still always an integral multiple of $hf/2e$. The factor is then at most about nm where m is the number of junctions. Because of the large quantisation steps $hf/2e$ occurring with the high frequencies f usually employed, the exact factor can easily be determined by a measurement of the total voltage with an ordinary digital voltmeter. By adjusting the d.c. current and the incident RF power, this factor can be externally set to the desired value.

Because of the large capacitance of the Josephson tunnel junctions usually used one can assume that the RF and d.c. voltages are impressed, conforming to (3.33). According to (3.36) the step n is of maximal width, and hence $2\Delta I_n$ is greatest, when the contribution of the Bessel function J_n is greatest, i.e.

$$n\frac{\hat{U}_1}{U_n} = x'_{n1} \tag{4.3}$$

where x'_{n1} is the location of the first zero of $\mathrm{d}J_n(x)/\mathrm{d}x$. For large n we have $x'_{n1} \simeq n$ and from (4.3) it follows that $\hat{U}_1 \simeq U_n$. For the peak values of $u(t)$ from (3.33) not to exceed the energy gap voltage U_{gap} we must have

$$U_n < \frac{U_{\mathrm{gap}}}{2} = \frac{\Delta}{e} \quad . \tag{4.4}$$

Otherwise rectification effects of the nonlinear quasi-particle tunnelling curve would according to Fig. 2.3 cause an additional direct current, which would move the microwave-induced steps in Fig. 3.10b still further away from the voltage axis than is already done below the energy gap voltage by the assumed linear mean resistance R_{sg}. Equation (4.4) represents an upper limit in principle of the zero-current steps. R_{sg} should be large, in order to approach this limit.

Under the condition (4.4), the quasi-particle tunnelling curve acts on the d.c. and a.c. voltage of the injected frequency f approximately as a linear resistance R_{sg} (see middle link of Fig. 4.1).

When operating with $I = 0$, the received RF power $\hat{U}_1^2 G_n/2$ generates d.c. power U_n^2/R_{sg} in the $\sin\varphi$ term. With $\hat{U}_1 \simeq U_n$ it follows from the power balance that $G_n \simeq 2/R_{\mathrm{sg}}$. Here $G_n = \mathrm{Re}(Y_n)$, where Y_n is the admittance offered by the $\sin\varphi$ term to the radio frequency oscillation with frequency f (see right-hand link of Fig. 4.1). The real part of the input admittance Y_{J} of the

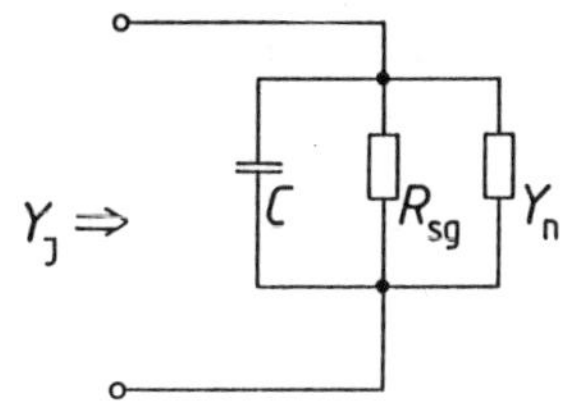

Fig. 4.1. Radio frequency equivalent circuit of the Josephson tunnel junction

Josephson tunnel junction is accordingly under these conditions equal to $3/R_{sg}$. Since $|\mathrm{Im}(Y_n)| \ll \omega C$, the imaginary part is determined almost entirely by the parallel plate capacitance $C = A\varepsilon_0\varepsilon_r/d$ (see left-hand link of Fig. 4.1). Here $A = l \cdot w$ is the area of the junction. Combining these results we find that

$$Y_J = \frac{3}{R_{sg}} + j\omega C \quad . \tag{4.5}$$

In order to be able to evaluate (4.5) and design the microwave circuit, we must first determine the critical current density J_c together with the dimensions l and w. The resistance R_{sg} is then found from J_c.

On the one hand J_c should be large, in order to obtain a large step width $2\Delta I_n$. However, J_c should not be so large that the plasma frequency comes close to the injection frequency, causing the steps to be distorted by chaos (see Sect. 3.2.2). For the sake of a large step width one also wants as large an area $A = lw$ as possible. The lateral dimension w is limited, however, by disturbing resonance effects [4.3], which arise if w is equal to or greater than a half wavelength in the Josephson tunnel junction. It therefore follows, with (3.41), that

$$w_{max} \leq \frac{v}{2f} = \frac{\sqrt{d/d'}}{2f\sqrt{\mu_0\varepsilon_0\varepsilon_r}} \quad . \tag{4.6}$$

If the length l increases beyond a limit l_{max} the Josephson tunnel current density becomes dependent on z. This means that for a length $l = l_{max}$ the step width $2\Delta I_n$ reaches saturation [4.4]. Larger values of l are not sensible. According to [4.5] one has

$$l_{max} = 3\lambda_J\sqrt{\frac{J_0 d}{2\pi f\varepsilon_0\varepsilon_r U_n}} = \frac{3}{\sqrt{n}}\,\frac{\sqrt{d/d'}}{2\pi f\sqrt{\mu_0\varepsilon_0\varepsilon_r}} \tag{4.7}$$

with λ_J from (3.44). For the critical current $I_c = J_c l_{max} w_{max}$ one finds, with J_c from (3.38), that

$$I_c = \frac{3}{2}\left(\frac{f_p}{f}\right)^2 \frac{\Phi_0}{d'\sqrt{n}\mu_0} \quad . \tag{4.8}$$

If the ratio $f/f_p \geq 4$ the steps are free from chaos, whereas if $3 \leq f/f_p \leq 4$ this is true only for sufficiently large RF amplitudes [4.3]. Since $d' \simeq 2\lambda$, materials with small penetration depths λ should be used for large critical currents and hence for large step widths. From (4.8), for example, for materials with d' =

274 nm, such as some lead alloys [4.6], together with $n = 6$ and the chaos-free maximum possible ratio $f_p/f = 1/4$, the critical current is $I_c = 230\mu$A. For elements whose insulating layer is formed by oxidation of the Pb-In-Au base electrode, the capacitance per unit area is $\varepsilon_0\varepsilon_r/d = 3\,\mu$F/cm^2 [4.7], almost independently of the critical current density. In the example, (4.6) and (4.7) show that, with an operating frequency of 70 GHz, the dimensions are $w_{max} = 70\mu$m and $l_{max} = 27\,\mu$m, with the critical current density $J_c = I_c/(w_{max}l_{max}) = 12\,$A/cm^2, which is to be realised by an appropriate method of manufacture (see Sect. 5.1.2). From (3.16) the product of critical current and normal resistance is

$$I_c R_n = \mathrm{SCF}\frac{\pi\Delta(T)}{2e}\tanh\frac{\Delta(T)}{2k_B T} \quad , \tag{4.9}$$

a quantity independent of the area A and of the critical current density J_c. For lead with SCF = 0.78 and $T = 4.2\,$K we accordingly have $I_c R_n = 1.3\,$mV. In the example, the normal resistance is thus given as $R_n = 5.6\ \Omega$. The ratio of the resistance R_{sg} below the energy gap voltage to the normal resistance R_n, depends strongly on the material and on the method of manufacture. For junctions on a Pb-In-Au base electrode a typical value is $R_{sg}/R_n \simeq 10$. Accordingly the equivalent circuit 4.1 has to take account of a resistance $R_{sg} \simeq 60\,\Omega$. The capacitance is given by the parallel plate condenser formula as $C = A\varepsilon_0\varepsilon_r/d = 56$ pF. The RF admittance of the Josephson tunnel junction according to (4.5) is therefore $Y_J = 24.6 \times (7 \times 10^{-4} + \mathrm{j}1)$S. It is quite predominantly capacitative. — Now we have described the individual Josephson tunnel junctions.

Figure 4.2a shows a section of the Josephson tunnel junctions connected in series [4.8]. On a non-conducting substrate, first of all a continuous superconducting layer S1 is deposited, forming the ground metallisation of the microstrip line circuit constructed thereon. The insulating layer I1 is its dielectric. The strip N of normally conducting material built on it acts as terminal resistance (see below). In the train of the series connection the superconducting layer S2 acts as a strip of the microstrip line. At the same time it forms the base electrode of the Josephson tunnel junctions. Above it is the insulating layer I2 with openings in it. In the region of these windows the base electrode is oxidised, so that the tunnel junctions are formed here after deposition of the top electrode S3.

Figure 4.2b is a rough RF equivalent circuit diagram, which however conveys the basic conception of the arrangement. A more detailed treatment is to be found in [4.4]. The connection in series of the Josephson tunnel junctions thus represents a periodic circuit, which can be calculated according to [4.9]. Because of the predominantly capacitative nature of Y_J it has basically high pass behaviour with the cut off frequency

$$f_{cK} \simeq \frac{1}{2\pi}\left(\frac{Z_0 l_K\sqrt{\mu_0\varepsilon_0\varepsilon_{r1}}}{2}\right)^{-1/2} \quad . \tag{4.10}$$

Here

$$Z_0 = \sqrt{\frac{\mu_0}{\varepsilon_0\varepsilon_{r1}}}\frac{d_1}{w_2} \tag{4.11}$$

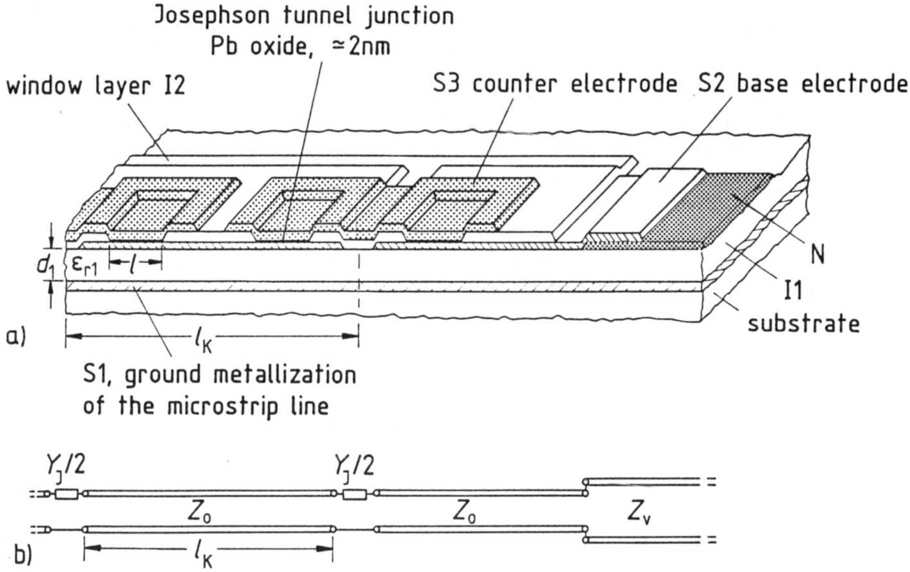

Fig. 4.2. Connection in series of Josephson tunnel junctions with terminal resistance. (**a**) Longitudinal section. Materials according to [4.8]: substrate: glass 0.3 mm; superconducting layers: S1 Nb 300 nm, S2 PbInAu 200 nm, S3 PbAu 250 nm; insulating layers I1 SiO 100 nm, I2 SiO 250 nm, together with tunnelling layer Pb oxide $\simeq$ 2 nm; resistance layer N InAu 250 nm, (**b**) RF equivalent circuit with periodic circuit structure

is the wave impedance of the unloaded microstrip line, whose dielectric I1 has thickness d_1 and relative permittivity ε_{r1} and whose strip has the width w_2 of the base electrode. The wave impedance of the periodic circuit is

$$Z_K = Z_0\sqrt{1-(f_{cK}/f)^2} \quad . \tag{4.12}$$

For $\varepsilon_{r1} = 5.7$ (SiO), $d_1 = 1\mu\text{m}, w_2 = 70\mu\text{m}, l_K = 100\mu\text{m}$ and with $C = 56$ pF and $f = 70$ GHz, we find that $Z_0 = 2.26\,\Omega$, $f_{cK} = 22.4$ GHz and $Z_K = 2.14\,\Omega \simeq Z_0$.

Just as the wave impedance Z_K of the periodic circuit when $f \gg f_{cK}$ is approximately equal to the wave impedance Z_0 of the unloaded microstrip transmission line, so also is the attenuation constant α_K of the periodic circuit then roughly equal to that of the unloaded microstrip line. It can be calculated from the known material constants by the methods shown in Sect. 1.4. For the example we can indeed estimate it directly from Fig. 1.16 if we assume that the attenuation constants of an antipodal fin-line with considerable overlap and a microstrip transmission line with the same dimensions of the overlapping region are equally large. According to [4.9] the attenuation constant of transmission lines with small resistance R' per unit length is given by

$$\alpha = \frac{R'}{2Z_0} \quad . \tag{4.13}$$

In microstrip lines the connection with the surface resistance R_S is approximately $R' \simeq 2R_S/w$. With (4.11) we then find that

$$\alpha \simeq \frac{R_S}{h}\sqrt{\frac{\varepsilon_0\varepsilon_r}{\mu_0}} \tag{4.14}$$

which is independent of the strip width w. Taking account of the roughly quadratic frequency dependence of R_S we obtain from the value $\alpha = 17\,\text{dB/m}$, measured from Fig. 1.16 for 30.4 GHz and $h = 0.65\,\mu\text{m}$ at $T = 4.2\,\text{K}$, the attenuation $\alpha_K = 0.06\,\text{dB/mm}$ for $f = 70\,\text{GHz}$ and $h = d_1 = 1\,\mu\text{m}$. In fact the attenuation will be a little larger still, since losses occur in the transmission line sections formed by the layers S2 and S3, as well as losses in the resistive component of Y_J.

The attenuation in the periodic circuit restricts the maximal number of Josephson junctions which can be directly connected in series, since the RF voltage over the individual junctions, which decreases from junction to junction as the wave propagates along the periodic circuit, must not change too much. Otherwise the state of maximal step width $2\Delta I_n$ would not be achieved for all junctions simultaneously. Then some elements would be operating with a narrower step than others, and the stability of the output voltage would thereby be reduced. Additionally, for RF voltages which are too small and for ratios $f/f_p \leq 4$, chaotic instability problems occur.

It has been established that, in designing a voltage standard, the RF voltage drops across the junctions along a chain may deviate from the optimal value by at most $\pm\delta$, where $\delta = 20\,\%$. The incident RF power is then chosen so that this optimal voltage $\hat{U}_1 = x'_{n1}U_n/n$ occurs at an element about in the middle of the chain, see (4.3).

The RF voltage varies along the periodic circuit because of its attenuation, but also because of reflections at a terminal impedance which is perhaps not ideally matched. If the circuit is designed on the principle that both effects should be equal, then a voltage variation of only $\pm\delta/2$ can be allowed for the effect of the attenuation alone. If one reckons that two Josephson junctions are included in each period length l_K, then the maximal number of junctions in a chain is

$$m_{\max} = -4\frac{\ln(1-\delta/2)}{\alpha_K l_K} \quad . \tag{4.15}$$

In the example this gives $m_{\max} = 610$. It must still be ensured, however, that the terminal impedance is matched sufficiently well. As it is difficult, with high frequencies f and low temperatures, to construct a concentrated terminal impedance sufficiently acccurately, one normally chooses a lossy open running transmission line of length l_v at the end as a load. According to [4.9], with sufficiently small longitudinal resistance per unit length R'_v, this transmission line has the input impedance

$$Z_v = \left[Z_{v0} - \mathrm{j}\frac{R'_v}{2\beta}\right]\frac{1+\mathrm{j}\tanh(\alpha_v l_v)\tan(\beta l_v)}{\tanh(\alpha_v l_v)+\mathrm{j}\tan(\beta l_v)} \tag{4.16}$$

with

$$\beta = \omega\sqrt{\mu_0\varepsilon_0\varepsilon_r}\,; \quad \alpha_v = \frac{R'_v}{2Z_{v0}}\,; \quad Z_{v0} = \sqrt{\frac{L'_v}{C'_v}} \quad . \tag{4.17}$$

Here L'_v and C'_v are the inductance and the capacitance per unit length.

According to (4.16) there are two causes for the input impedance to differ from the value Z_{v0}: because of the losses in R'_v the wave impedance is no longer real, and, because of the finite length l_v of the transmission line reflections from the open end can react on the input. The terminating transmission line is first of all now adjusted in its cross-sectional geometry so that $Z_{v0} = Z_K$. An approximate prolongation of the geometry of the periodic circuit structure is therefore advisable (see layer N in Fig. 4.2).

Then R'_v and l_v are set so that the maximal possible mismatches from both causes are equal, and a total return loss of $-20 \log \delta/2$ is not exceeded in the most unfavourable case. The required longitudinal resistance per unit length is achieved by the choice of convenient material and thickness for the normally conducting layer N. Alloys of In and Au have proved to be a good choice. In [4.10] it is shown how their resistance per unit area R_A is related to the layer thickness. The resistance per unit area $R_A \simeq R'_v w_v$ is related to the longitudinal resistance per unit length by the width w_v of the normally conducting strip.

The quantised total voltage which one can draw from a chain with a terminal impedance is $nm_{max}hf/2e$. If the output voltage of the standard needs to be larger, several chains must be connected in series with respect to d.c.; and if they need to be operated from only one microwave source, they must be connected in parallel with respect to RF.

Figure 4.3 shows the layout of a chip with four such chains, folded back on themselves in order to get the Josephson junctions D on the smallest possible, roughly square, surface [4.11,12]. Then the sample variation in the manufacture of the junctions is also small. The chains receive their RF excitation from one waveguide, which has longitudinal slits in the broad walls at the open end. The chip with the fin-line taper B [4.13, 14] is placed in these slits. The incident TE_{10} mode in the waveguide is thus transformed to the fin-line wave and from there to a wave on the microstrip transmission line. At the microstrip "T" junction two DC-blocks C are attached, which consist of a $\lambda/4$ overlap of two strips insulated from each other. At two further "T" junctions the incident RF power is fed into the chains. At the ends of the chains are the absorbing transmission lines F, and the direct current connection A is made via RF band stop filters E. In the region of the microstrip circuit the metal ground plane S1 (see Fig. 4.2a) is reduced so much that it does not have a strong RF effect. In this way one can avoid the magnetic flux being frozen in when cooling, at faults in the layer S1 which act as pinning centres. This could also occur, if S1 is fault-free but consists of a hard superconductor of the second class, see Sect. 1.6. The magnetic fields of such captured fluxes would lower the critical currents I_c of the Josephson junctions and also significantly reduce the step widths $2\Delta I_n$.

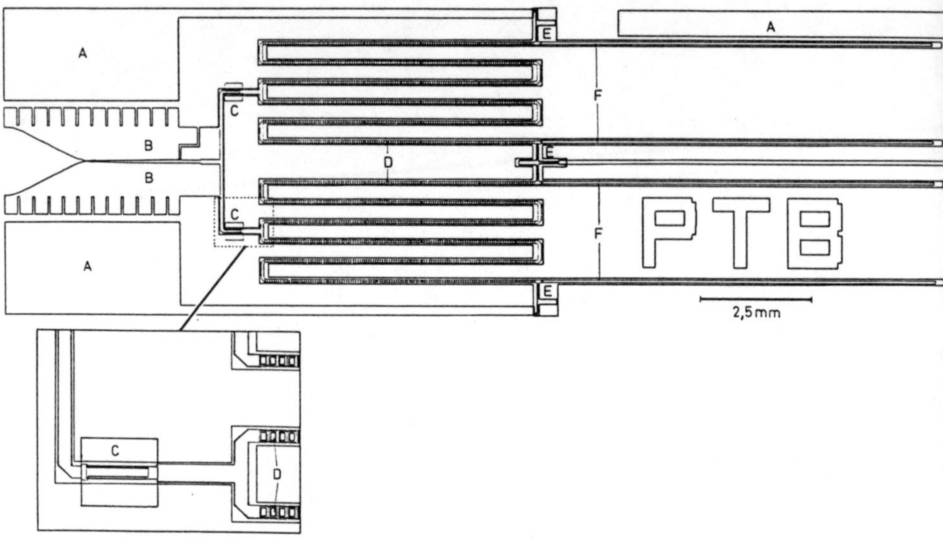

Fig. 4.3. Layout of a chip for a voltage standard with 1440 Josephson tunnel junctions, from [4.11]. A – d.c. voltage connection, B – fin-line taper, C – DC-blocks, D – Josephson tunnel junctions, E – band stop filters, F – absorbing transmission lines

The circuit of Fig. 4.3 makes it possible to provide quantised voltages between 0.1 and 1.3 V which are stable for more than five hours [4.12]. A measure of the accuracy of the voltage is its deviation ΔU from the horizontal along the steps. With an output voltage of 1 V, ΔU is less than 10^{-9} V. This limit is imposed by the intrinsic noise of the measuring device at room temperature [4.12]. ΔU can be controlled to a resolution of 7×10^{-13} V by comparing the output voltages of two chips as in Fig. 4.3 at low temperatures by means of a SQUID (see Sect. 3.4). Even with this comparison, no ΔU could be measured within this resolution [4.15].

Circuits with Josephson junctions of lead alloys can only be stored for a limited time at ambient temperature, and can be cooled from room temperature to low temperature only on a few occasions without their properties significantly deteriorating (see Sect. 5.1.1). Such chips for Josephson voltage standards are accordingly best made out of more stable materials. Examples are circuits with Josephson tunnel junctions made from Nb/Al-Oxide/Nb and NbN/MgO/NbN [4.3]. A chip of the described type for a Josephson voltage standard, containing 14 184 junctions and providing standard voltages of > 10 V is described in [4.59].

4.2 Detectors

Josephson junctions which are used as detectors for very weak signals should have the smallest possible parallel capacitance. Point contacts and microbridges accordingly come up for consideration. They can be described approximately by the RSJ model. The RF source is often modelled in the equivalent circuit diagram by an ideal a.c. current source $\hat{I}_S \sin \omega_S t$, which lies parallel to the direct current source I (Fig. 4.4). For weak microwave injection the d.c. curve is shaped as in Fig. 4.5. We see an operating state in which the first step begins to originate. From the model of impressed voltage we can derive the difference ΔI by which at constant small operating voltages $U \to 0$ the current I is diminished because of the RF current $\hat{I}_S \sin \omega_S t$. Equation (3.36) gives for the zeroth step, i.e. with $n = 0$ and a development of the Bessel function for small arguments

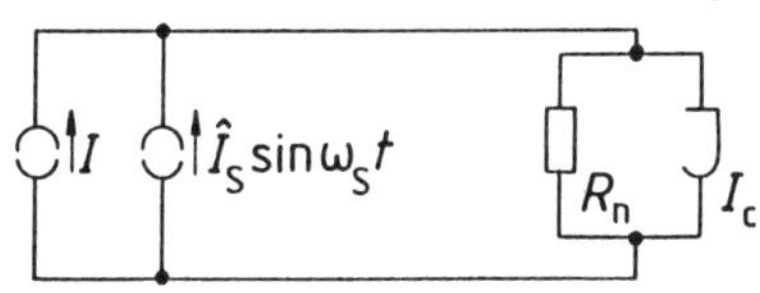

Fig. 4.4. Equivalent circuit for the Josephson junction in the RSJ model with impressed d.c. and a.c. current

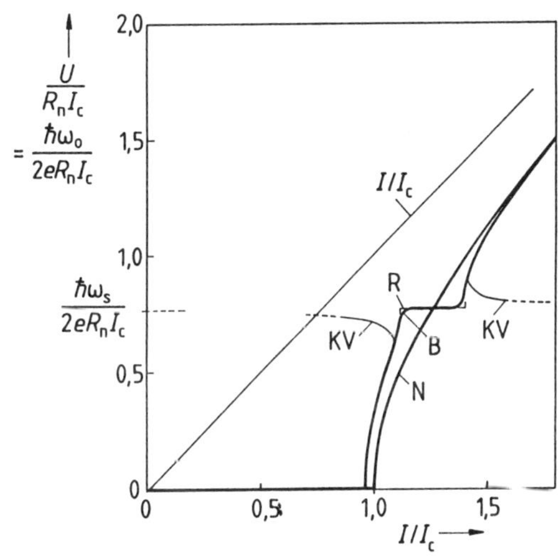

Fig. 4.5. Current-voltage characteristics of the Josephson junction in the RSJ model. (N) $\hat{I}_1 = 0$. Otherwise $\hat{I}_1/I_c = 0.3$; $\xi = \omega_S/\omega_c = 0.77$; (KV) from [4.16]; (R) from [4.17]; (B) observed behaviour rounded by noise

$$\Delta I = I_c \left[J_0(0) - J_0\left(\frac{2e\hat{U}_S}{\hbar\omega_S}\right)\right] \simeq I_c \frac{\hat{U}_S^2 e^2}{\hbar^2 \omega_S^2} \quad . \tag{4.18}$$

Here $\hat{U}_S$ is the RF voltage with the angular frequency ω_S at the junction. At operating points with $\omega_0 = 2eU/\hbar \ll \omega_S$, according to [4.17] the input impedance of the parallel connection of R_n and J in Fig. 4.4 is $\hat{U}_S/\hat{I}_S = R_n$. From (4.18) we therefore find that

$$\Delta I = \frac{\hat{I}_S^2}{4I_c}\frac{\omega_c^2}{\omega_S^2} \quad , \tag{4.19}$$

where ω_c is substituted from (3.32). For operating points $\omega_0 > 0$ and for the case of sufficiently small amplitude $\hat{I}_S$, ΔI can be calculated as in Sect. 3.2.1. Instead of (3.28) we now obtain for the total current

$$I + \hat{I}_S \sin \omega_S t = I_c \sin \varphi + \frac{1}{R_n}\frac{\hbar}{2e}\frac{d\varphi}{dt} \quad . \tag{4.20}$$

When $\hat{I}_S \ll I_c$ this equation gives approximately, according to [4.16],

$$\Delta I = \frac{\hat{I}_S^2}{4I}\frac{\omega_c^2}{\omega_S^2 - \omega_0^2} \, , \quad \omega_S \neq \omega_0 \quad . \tag{4.21}$$

The direct current curve from (4.21) is plotted in Fig. 4.5 as curve KV. According to whether the operating point ω_0 is smaller or larger than ω_S, ΔI is positive or negative. Equation (4.21) is not valid for $\omega_0 \simeq \omega_S$. Such operating points lie in the region of the first microwave-induced step. With the step width ΔI_1 from Fig. 3.12 the behaviour in this region is given by the curve R in Fig. 4.15. Usually the step is rounded by noise [4.18], so that it can appear as the curve B.

At operating points $\omega_0 \ll \omega_s$ and $\omega_0 \gg \omega_S$ one can make use of the distortion of the curves for broad band detection of microwave signals. We shall explain this operation in the next section. When $\omega_0 \simeq \omega_S$, in contrast to the usual quadratic detectors, the curves contain not only information on the amplitude of the incident signal, but also on its frequency. We shall call detectors which make use of this phenomenon spectral detectors and treat them in Sect. 4.2.2.

4.2.1 Broad Band Detectors

Equation (4.21) describes a quadratic detector. The output amplitude ΔI is proportional to the square $\hat{I}_S^2$ of the input amplitude. The Josephson junction is there operated normally with impressed direct current. One then measures the voltage drop $\Delta U = \Delta I R_d$ varying with the incident RF power, where R_d is the dynamic resistance of the undistorted curve according to (3.31), and hence

$$R_d = \frac{dU}{dI} = R_n \frac{I/I_c}{\sqrt{(I/I_c)^2 - 1}} \quad . \tag{4.22}$$

From the analysis of [4.16] we find that the real part R_{in} of the RF input impedance is

$$R_{in} = \begin{cases} R_n & \text{for } \omega_0 \ll \omega_S \\ R_d & \text{for } \omega_0 \gg \omega_S \end{cases} \quad . \tag{4.23}$$

So the voltage sensitivity of the detector is then

$$R_u = \frac{\Delta U}{P_{in}} = \frac{2\Delta I R_d}{\hat{I}_S^2 R_{in}} \tag{4.24}$$

from (4.21–23) and, with $\omega_0/2\pi = 2eU/h = 2eR_nI_c\sqrt{(I/I_c)^2 - 1}/h$, it becomes

$$R_u = \frac{\omega_c^2}{2I\omega_S^2}\frac{R_d}{R_n} \quad \text{for} \quad \omega_0 \ll \omega_S \tag{4.25}$$

$$R_u = \frac{1}{2R_d}\frac{d^2U}{dI^2} \quad \text{for} \quad \omega_0 \gg \omega_S \quad . \tag{4.26}$$

Equation (4.26) describes the classical quadratic detector, whose sensitivity is proportional to the curvature of the characteristic curve. The true Josephson detector, however, uses operating points and voltage sensitivities according to (4.25). With low frequencies ω_S, R_u is large for high characteristic frequencies ω_c and hence for high values of the product R_nI_c. Moreover, R_u increases with R_d according to (4.22) beyond all limits if I lies only slightly above I_c at the operating point. In practice, however, R_d is limited by the rounding of the curve by noise (see Fig. 3.6).

The intrinsic noise also limits the minimal detectable power of the detector. The noise equivalent power NEP is calculated as that RF power which causes a current $|\Delta I|$ that is just as great as the effective noise current. If white thermal noise is assumed, this is equal to

$$\sqrt{i_N^2} = \sqrt{\frac{4k_BTB}{R_n}} \quad . \tag{4.27}$$

Here B is the recording band width of the detector. Using (4.27), the NEP of the Josephson detector ($\omega_0 \ll \omega_S$), assuming impedance match, is found to be

$$\text{NEP} = \frac{4I\sqrt{k_BTBR_n}\omega_S^2}{\omega_c^2} \quad . \tag{4.28}$$

For operating points with $eU \ll k_BT$ ($k_BT/e = 360\,\mu$V at 4 K), R_n in (4.27) is to be replaced by $R_0 = U/I$ according to [4.19]. Then the NEP becomes

$$\text{NEP} = \frac{4IR_n\sqrt{k_BTB/R_0}\omega_S^2}{\omega_c^2} \quad . \tag{4.29}$$

A typical experiment is described in [4.16]. A point contact was used in a waveguide with $I_c = 10\mu$A, $R_n = 40\,\Omega$, $I = 1.2I_c$, $U \simeq 10\mu$V, $T = 4$ K and $f = 90$ GHz. The calculated NEPs from (4.28) and (4.29) were 1×10^{-16}W/$\sqrt{\text{Hz}}$ and 8×10^{-16}W/$\sqrt{\text{Hz}}$, respectively. The measured NEP was 5×10^{-15}W/$\sqrt{\text{Hz}}$, a value about an order of magnitude higher than the calculated ones.

Also Josephson tunnel junctions have been successfully used as detectors [4.20]. Thorough analyses and optimisations of the circuits are to be found in [4.21] and [4.22]. Although the NEP of Josephson detectors is comparable with that of low-temperature bolometers, their extremely short response time especially could be advantageous in some applications [4.19].

4.2.2 Spectral Detectors

The spectral Josephson detector [4.23] gives information not only on the amplitude of an incoming signal, but also on its frequency. So in contrast to the mixer, no separate local oscillator is needed. The device is therefore extremely simple.

For operating points ω_0 lying close to ω_S, in the noise-free case the deformation ΔI of the d.c. characteristic (curve R in Fig. 4.5) depends linearly on the incoming microwave signal. This is the case because the step width ΔI_1 for small RF amplitudes is proportional to $\hat{U}_1$ from (3.36) in the model of the impressed RF voltage, just as it is proportional to $\hat{I}_1$ (Fig. 3.12) in the model of the impressed RF current. However, the intrinsic noise of the Josephson junction rounds off the step so much, according to [4.24] (see curve B in Fig. 4.5), that ΔI is proportional to the square of the RF amplitude. Under these conditions (4.21) has to be generalised in the following way [4.25,26], [4.24].

$$\Delta I = -\frac{\hat{I}_S^2 \omega_c^2}{8 I \omega_0}\left[\frac{\omega_0+\omega_S}{(\omega_0+\omega_S)^2+\delta^2}+\frac{\omega_0-\omega_S}{(\omega_0-\omega_S)^2+\delta^2}\right] \quad . \tag{4.30}$$

Here δ is the width of the spectral line of the Josephson oscillator determined by noise and in some cases by disturbing currents. Corresponding to the applied d.c. voltage U, a strong alternating current flows through the junction with the fundamental frequency $\omega_0/2\pi$. The spectral resolution of the spectral Josephson detector is therefore determined, according to (4.30), by the line width δ. The line broadening by noise occurs because the noise voltages with frequency components $\omega_N \ll \omega_0$ cause a frequency modulation of the Josephson oscillator and hence give rise to a frequency spectrum whose width $\delta \neq 0$.

For a Josephson junction that obeys the RSJ model, we can estimate δ in the following way, as shown in [4.27]. We assume that only white thermal noise of the normal resistance $R = R_n$ in Fig. 3.4 is to be taken into account. Then a noise current source is connected in parallel to the resistance R_n, having in the frequency region from ω_N to $\omega_N + \Delta\omega_N$ the root-mean-square (r.m.s.) value $\left(\Delta\overline{i_N^2}\right)^{1/2}$ where

$$\Delta\overline{i_N^2} = \frac{4 k_B T \Delta\omega_N}{2\pi R_n} \quad . \tag{4.31}$$

We shall see that only low frequencies $\omega_N \ll \omega_0$ contribute to the line broadening. Neglecting the parallel capacitance we find the low frequency noise voltage from $\Delta\overline{u_N^2} = \Delta\overline{i_N^2} R_d^2$ with R_d taken from (4.22). We now substitute $\Delta\overline{u_N^2}$ by a sinusoidal voltage of equal r.m.s. value at the frequency ω_N

$$U_1 = \hat{U}_1 \cos \omega_N t \quad \text{with} \quad \hat{U}_1 = \sqrt{2\Delta\overline{u_N^2}} \quad . \tag{4.32}$$

We calculate the spectrum given by this modulation voltage from the impressed voltage model. From (3.35), putting

$$x = \frac{2e\hat{U}_1}{\hbar\omega_N} \ll 1 \tag{4.33}$$

we obtain the current

$$\begin{aligned} i_J(t) =& I_c\Big\{\sin(\omega_0 t + \varphi_0) - \frac{x}{2}\sin\big[(\omega_0 - \omega_N)t + \varphi_0\big] \\ & + \frac{x}{2}\sin\big[(\omega_0 + \omega_N)t + \varphi_0\big]\Big\} \quad . \end{aligned} \tag{4.34}$$

We shall now take account of the RSJ model, in that we replace I_c in (4.34) by the amplitude $\hat{I}_0$ of the component of the current at ω_0 in the noise-free RSJ model. We could, if required, calculate $\hat{I}_0 = \hat{U}_0/R_n$ from the Fourier component $\hat{U}_0$ of the voltage (3.30) at the frequency ω_0.

For the carrier oscillation line this substitution of I_c by $\hat{I}_0$ is certainly correct. For the side bands at $\omega_0 \pm \omega_N$, however, it is also right, since one can cause this to happen by mixing with the carrier. They must therefore be proportional to the carrier amplitude. We thus obtain from (4.34) for the mean square value of the current in the side band from $\omega_0 \pm \omega_N$ to $\omega_0 \pm (\omega_N + \Delta\omega_N)$

$$S_i(\omega) = \frac{\hat{I}_0^2 x^2}{8} = 2\pi \hat{I}_0^2 \left(\frac{2e}{h}\right)^2 k_B T \frac{R_d^2}{R_n} \frac{1}{\omega_N^2} \quad . \tag{4.35}$$

With $\omega_N^2 = (\omega - \omega_0)^2$, (4.35) represents the flanks of a Lorentz curve for sufficiently large distance from the carrier. In the neighbourhood of the carrier (4.35) is not valid because of (4.33). We now assume that even in the immediate neighbourhood of the carrier the spectrum forms a Lorentz curve according to

$$S_{iL}(\omega) = P_i \frac{\delta/\pi}{\omega_N^2 + \delta^2} \tag{4.36}$$

with

$$P_i = \int_0^\infty S_{iL}(\omega) d\omega \quad . \tag{4.37}$$

In what follows we shall assume that the total power P_i normalised to the load resistance in the case with noise is equal to the power $P_i = \hat{I}_0^2/2$ in the noise-free case. Then by equating (4.35) and (4.36) with $\omega_N^2 \gg \delta^2$ we find the full line width of the Josephson oscillation to be

$$2\Delta f = \frac{2\delta}{2\pi} = 4\pi \left(\frac{2e}{h}\right)^2 k_B T \frac{R_d^2}{R_n} \quad . \tag{4.38}$$

Equation (4.38) agrees with a somewhat more accurate calculation, which does not rely on the impressed voltage model [4.27,26]. For the calculation of the line width of Josephson tunnel junctions we refer to [4.28]. Experimentally measured line widths are often larger than that given by (4.38). External noise which overlies the applied direct current is certainly often the cause. Increases in noise caused by chaos at Josephson tunnel junctions are described in [4.29].

The spectral Josephson detector finds a possible application as a spectrometer for incoherent spectra $S_i(\omega_S)$ which vary only slightly over frequency differences $\delta/2\pi$ [4.26, 24, 30]. According to [4.26] $\hat{I}_S^2/2$ can be replaced in (4.30) by $S_i(\omega_S)\mathrm{d}\omega_S$ and the right-hand side can be integrated over ω_S to find the distortion ΔI through all the spectral components of $S_i(\omega_S)$. As $\delta \to 0$ the result becomes

$$\Delta I = \frac{\pi\omega_c^2}{4I\omega_0}\mathcal{H}[S_i(\omega_S)] \quad . \tag{4.39}$$

Here $\mathcal{H}$ denotes the Hilbert transform

$$\mathcal{H}\left[S_i(\omega_S)\right] = \frac{1}{\pi}\int_{-\infty}^{\infty} \frac{S_i(\omega_S)}{\omega_S - \omega_0}\mathrm{d}\omega_S \quad . \tag{4.40}$$

The spectrum $S_i(\omega_S)$ is then obtained from the inverse transformation $\mathcal{H}^{-1}$

$$S_i(\omega_S) = \mathcal{H}^{-1}[g(\omega_0)] \tag{4.41}$$

with

$$g(\omega_0) = \frac{4I(\omega_0)\omega_0\Delta I(\omega_0)}{\pi\omega_c^2} \quad . \tag{4.42}$$

Because of the properties of the Hilbert transform, we have $\mathcal{H}^{-1} = -\mathcal{H}$, [4.31].

To evaluate (4.41) according to (4.42), we have first to establish the direct current curve with the spectrum switched off, and hence to determine I as a function of the normalised d.c. voltage $\omega_0 = 2eU/\hbar$. Then we have to measure the deformation $\Delta I(\omega_0)$ of the direct current curve by the incident spectrum $S_i(\omega_S)$. Now $g(\omega_0)$ is calculated according to (4.42) and the inverse Hilbert transformation carried out as in (4.41). The result is then the required spectrum $S_i(\omega_S)$.

Spectrometers working on this principle and whose Josephson junctions consist of niobium or lead alloys can be used between about 30 GHz and 1 THz. Their frequency resolution is a few hundred MHz in the region of $\omega_S/2\pi = 70$ GHz [4.30]. Referred to the same instantaneous reception band width their sensitivity is much higher than that of ordinary spectrum analysers. The evaluation of (4.42) and (4.41) is conveniently carried out on a computer. For more details on the spectral detector see [4.60].

4.3 Mixers

Because of their strong nonlinearity Josephson junctions can also be used as frequency converters. For this purpose point contacts have up till now been used almost exclusively. The mixing process can be explained using the RSJ model.

According to Fig. 4.5 the zero order step is somewhat suppressed by the injected LO power with the frequency ω and the first develops. Using a bias current I_{dc} the operating point of the mixer is set at about half-way between the two steps. Simplifying somewhat, the effect of an additional signal current with the frequency ω_S can be understood as if the local oscillator current i_H was amplitude modulated with the frequency $\omega_I = \omega_S - \omega$. The curve in Fig. 4.5 then swings as a whole in the horizontal direction, as the width I_0 of the zero order step varies with the amplitude I_H of the local oscillator current. To the d.c. voltage at the junction is thereby added an a.c. voltage with the frequency ω_I and with an amplitude proportional to $R_d(dI_0/dI_H)$. The conversion gain should then be proportional to $R_d(dI_0/dI_H)^2$.

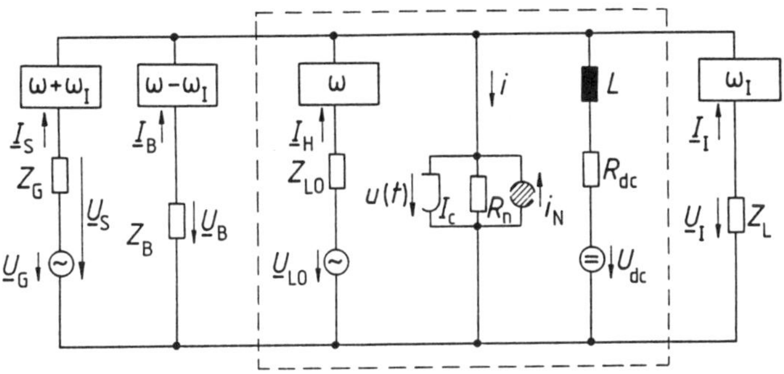

Fig. 4.6. Equivalent circuit of the Josephson mixer with current control

A detailed analysis of the Josephson mixer is illustrated by the equivalent circuit diagram in Fig. 4.6 from [4.32]. Only currents with the LO frequency ω, the signal frequency $\omega_S = \omega + \omega_I$, the image frequency $\omega_B = \omega - \omega_I$ and the intermediate frequency ω_I can flow through the Josephson junction. In addition there is a direct current produced by a voltage source U_{dc} with high internal resistance R_{dc} and a noise current i_N, which with $\overline{i_N^2} = 4k_BTB/R_n$ represents the thermal noise of the parallel resistance R_n. To calculate the conversion matrix of the portion of the circuit enclosed by the broken line, we start from the equations dual to (2.53 – 58). For this it is necessary first to study the Josephson junction without the small signal, i.e. with only the d.c. and LO oscillation. With the phasor $\underline{I}_H$ of the local oscillator current we then have

$$i = I_{dc} + \frac{1}{\sqrt{2}}\left(\underline{I}_H e^{j\omega t} + \underline{I}_H^* e^{-j\omega t}\right) \quad , \tag{4.43}$$

and with $u = (\hbar/2e)d\varphi/dt$ it follows that the total current at the Josephson junction is

$$\frac{\hbar}{2eR_n}\frac{d\varphi}{dt} + I_c \sin\varphi = I_{dc} + \frac{1}{\sqrt{2}}\left(\underline{I}_H e^{j\omega t} + \underline{I}_H^* e^{-j\omega t}\right) + i_N(t) \quad . \tag{4.44}$$

The resultant voltage $u(t)$ at the junction contains a d.c. voltage contribution U_{dc}, an a.c. voltage contribution $\underline{U}_H$ with the frequency ω together with higher harmonic frequency components. The latter are neglected, since they can cause no currents in the rest of the circuit. As the solution of (4.44) we obtain the functional relationships $U_{dc}(I_{dc}, |\underline{I}_H|)$ and $\underline{U}_H(I_{dc}, \underline{I}_H)$. Unfortunately no explicit formulations for these are known. In [4.32] equation (4.44) was therefore integrated numerically and the voltage $u(t)$ subsequently submitted to Fourier analysis. The noise source $i_N(t)$ was simulated during the integration by a random number generator. Figure 4.7 shows a typical result for real $\underline{I}_H = \sqrt{2} \times 0.45 I_c$ and a normalised frequency $\Omega_H = \hbar\omega/(2eR_n I_c) = 0.4$. The noise parameter

$$\Gamma = \frac{2ek_B T}{\hbar I_c} \tag{4.45}$$

in Fig. 4.7 is equal to 0.01. It is clearly seen how the microwave-induced steps are rounded off by noise. The RF input impedance $\underline{U}_H/\underline{I}_H$ is predominantly real for $I_{dc}/I_c > 0.5$ and tends to R_n for large I_{dc}.

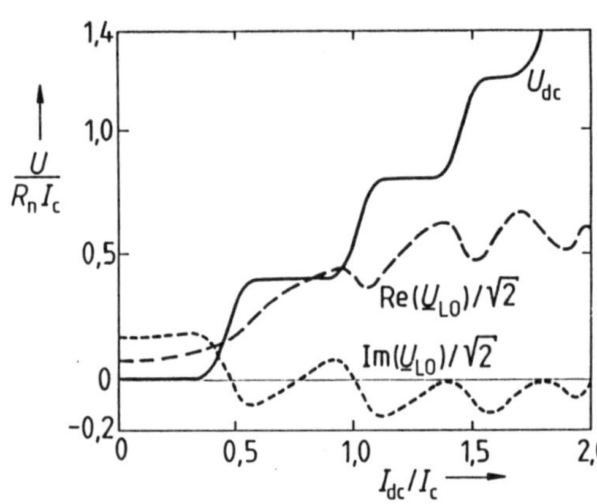

Fig. 4.7. d.c. voltage U_{dc} and phasor $\underline{U}_H$ of the voltage component at the frequency ω as functions of the direct current I_{dc}. $\underline{I}_H = \sqrt{2} \times 0.45 I_c$; $\Omega_H = \hbar/2eR_n I_c = 0.4$; $\Gamma = 2ek_B T/\hbar I_c = 0.01$. From [4.32]

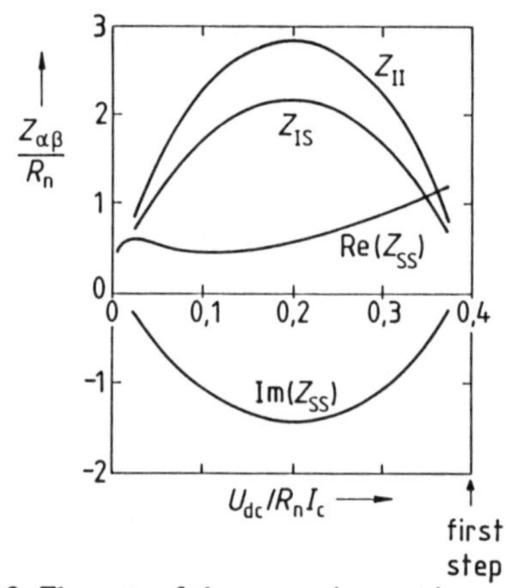

Fig. 4.8. Elements of the conversion matrix as functions of the d.c. voltage between the steps of zero and first order. Calculated for values of $\underline{I}_H, \Omega_H$ and Γ as in Fig. 4.7. From [4.32]

From the direct current characteristic curve $U_{dc}(I_{dc}, |\underline{I}_H|)$ and the RF input resistance $\underline{U}_H(I_{dc}, \underline{I}_H)/\underline{I}_H$ for LO level we can now, from the equations dual to (2.53 – 58), calculate the elements of the conversion matrix, which relate the current and voltage phasors at the signal, intermediate and image frequencies

$$\begin{bmatrix} \underline{U}_S \\ \underline{U}_I \\ \underline{U}_B^* \end{bmatrix} = \begin{bmatrix} Z_{SS} & Z_{SI} & Z_{SB} \\ Z_{IS} & Z_{II} & Z_{IB} \\ Z_{BS} & Z_{BI} & Z_{BB} \end{bmatrix} \begin{bmatrix} \underline{I}_S \\ \underline{I}_I \\ \underline{I}_B^* \end{bmatrix} , \quad \underline{U} = \underline{Z}\,\underline{I} . \tag{4.46}$$

Figure 4.8 shows the behaviour in this example of $\mathrm{Re}(Z_{SS})$, $\mathrm{Im}(Z_{SS})$, Z_{IS} and Z_{II} as functions of the d.c. voltage U_{dc} at the operating point. Apart from $\mathrm{Re}(Z_{SS})$, all quantities show symmetrical behaviour between the steps of the zero and first order. For maximal Z_{IS} the operating point of the mixer should be set at a point half-way between these two steps.

Since the exact form of the d.c. characteristic with microwave-induced steps depends strongly on the noise parameter Γ, the elements of the conversion matrix are also strongly influenced by Γ. For $\Gamma < 0.1$ the maximal values of Z_{IS} and Z_{II} relative to variations in U_{dc} are proportional to $1/\sqrt{\Gamma}$. For $\Gamma \to 1$, because of extreme rounding of the steps, Z_{IS} is very small and a downward mixing is no longer effective.

In order to calculate the conversion gain, the external circuit lying outside the broken line in Fig. 4.6 is here, just as in Sect. 2.3.1, described by a matrix, namely $\boldsymbol{Z}_e$:

$$\begin{bmatrix} \underline{U}_S \\ \underline{U}_I \\ \underline{U}_B^* \end{bmatrix} + \begin{bmatrix} Z_G & 0 & 0 \\ 0 & Z_L & 0 \\ 0 & 0 & Z_B^* \end{bmatrix} \begin{bmatrix} \underline{I}_S \\ \underline{I}_I \\ \underline{I}_B^* \end{bmatrix} = \begin{bmatrix} \underline{U}_G \\ 0 \\ 0 \end{bmatrix} \tag{4.47}$$

$$\underline{\boldsymbol{U}} + \boldsymbol{Z}_e \underline{\boldsymbol{I}} = \underline{\boldsymbol{U}}_G \quad . \tag{4.48}$$

Equations (4.46) and (4.47) can be solved for small signal currents

$$\underline{\boldsymbol{I}} = \boldsymbol{Y}' \underline{\boldsymbol{U}}_G \tag{4.49}$$

with

$$\boldsymbol{Y}' = (\boldsymbol{Z} + \boldsymbol{Z}_e)^{-1} = \begin{bmatrix} Y'_{SS} & Y'_{SI} & Y'_{SB} \\ Y'_{IS} & Y'_{II} & Y'_{IB} \\ Y'_{BS} & Y'_{BI} & Y'_{BB} \end{bmatrix} \quad . \tag{4.50}$$

The conversion gain G as the ratio of the real power applied to the load to the power available from the generator follows from (4.47) as

$$G = \frac{|\underline{I}_I|^2 \mathrm{Re}(Z_L)}{|\underline{U}_G|^2 / 4\mathrm{Re}(Z_G)} = 4|Y'_{IS}|^2 \mathrm{Re}(Z_L)\mathrm{Re}(Z_G) \quad . \tag{4.51}$$

For the assumed operating point U_{dc}, $|\underline{I}_H|$ we can now maximise G over variations of Z_G and Z_B. The load impedance Z_L has to be set for power maximisation according to the conjugate complex of the output impedance of the mixer

$$Z_{out} = Y'^{-1}_{II} - Z_L \quad . \tag{4.52}$$

This is valid at least so long as Re $(Z_{out}) > 0$. Operating states of the Josephson mixer with Re $(Z_{out}) < 0$ are indeed possible in principle. The mixer noise temperature, however, is then inconveniently high.

The noise current source $i_N(t)$ in Fig. 4.6 causes not only a deterioration in the coefficients in the conversion matrix: the noise currents from $i_N(t)$ at various frequencies are also transformed by mixing to the intermediate frequency. This effect is included in the numerical analysis of [4.32], so that the noise temperature T_M of the mixer can be calculated.

Figure 4.9 displays curves of constant noise temperature for a fixed operating point $U_{dc}/R_n I_c = 0.16, I_H = 0.45\sqrt{2} I_c$ in the complex plane of the generator

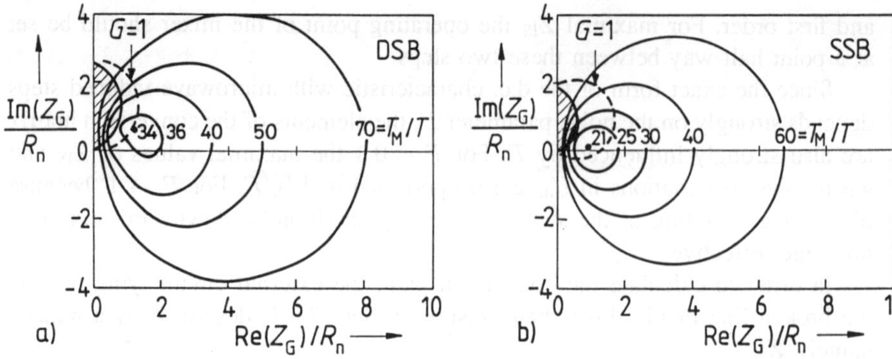

Fig. 4.9. Ratio of mixer noise temperature T_M to the temperature T of the Josephson junction as a function of the generator internal impedance Z_G for fixed operating point voltage $U_{dc}/R_n I_c = 0.16$. Calculated for values of $\underline{I}_H$, Ω_H and Γ as in Fig. 4.7. The points indicate impedances Z_G for minimal noise temperatures. In the shaded regions $\mathrm{Re}(Z_{out}) < 0$. **(a)** DSB: double side band mixer with $Z_B = Z_G$, **(b)** SSB: single side band mixer with $Z_B = 4jR_n$. From [4.32]

internal impedance Z_G, with fixed parameters $\Omega_H = 0.4$ and $\Gamma = 0.01$. The points in Fig. 4.9 describe Z_G for minimal T_M, i.e. $T_{MDSB}/T = 34$ and $T_{MSSB}/T = 21$. In the calculation for the double side band mixer (DSB) $Z_G = Z_B$ is assumed. For the single side band mixer (SSB) $Z_B = 4jR_n$ (hence practically an open circuit) is found to be optimal. This value is taken as the basis of Fig. 4.9b.

The broken curves in Fig. 4.9 separate the regions with $G > 1$ and $G < 1$. In the shaded regions $\mathrm{Re}(Z_{out}) < 0$. There G formally goes to ∞, but the mixer operation is unstable. Stable operation with amplification $G > 1$ is thus possible in the lunar-shaped regions, inside which and on the edges of which, respectively, lie the points of minimal noise temperature.

Figure 4.10 shows the construction of an experimental Josephson mixer with Nb point contact for $f_S = 36\,\mathrm{GHz}$ from [4.33]. The normal resistance of the Josephson junction is set at room temperature by turning the Nb screw. It then varies only slightly on cooling to low temperatures. The tuning plug and the adjustable waveguide short circuit can be varied continuously at low temperatures to optimal mixer behaviour. Grooves in the Nb flange form a band stop filter for the high frequencies. The intermediate frequency is transmitted over a coaxial cable. Experimental Josephson mixers, [4.33–35], show noise temperatures of $T_M/T = 20 \ldots 50$ and conversion gains of $0.5 \ldots 1.3$ at normalised frequencies $\Omega_H = 0.3 \ldots 0.4$ and noise parameters $\Gamma \leq 0.01$. The agreement with calculated results is good.

The noise temperature of SIS mixers is lower in the millimetre wave region than that of Josephson mixers. They are therefore to be preferred in this wavelength region. Since SIS mixers suffer from the Josephson effect in the submillimetre wave region, however, Josephson mixers could find their application here.

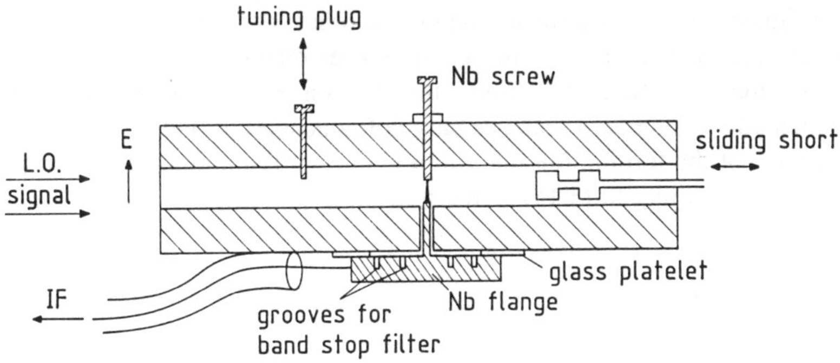

Fig. 4.10. Josephson mixer, from [4.33]

Besides the ability of the Josephson mixer to operate with an external local oscillator, it appears especially attractive to make use of the internal oscillations caused by the a.c. Josephson effect instead of an external local oscillator. One could in this way construct relatively easily a receiver, rapidly tunable over a wide frequency range. A serious problem arises, however, from the quite large line width of the Josephson oscillation, see (4.38), which often makes the frequency resolution of the receiver unacceptable. In [4.36] a Josephson mixer is described with internal LO which uses a niobium point contact. It was tested between 0.6 and 2 THz. Below the energy gap frequency $f_{gap} = 4\Delta/h = 1.4$ THz the mixer behaved as expected; above it the noise temperature increased drastically, in fact proportionally to f^6.

Josephson junctions have also been considered as harmonic mixers for the precise measurement of the frequencies of laser lines up into the terahertz region [4.38, 37].

4.4 Amplifiers

The $\sin\varphi$ term of a Josephson junction can be understood as a loss-free, nonlinear inductance. Then from $u = (\hbar/2e)\mathrm{d}\varphi/\mathrm{d}t$ and $i = I_c \sin\varphi$ we obtain

$$u = \frac{\hbar}{2e}\,\frac{1}{I_c\cos\varphi}\,\frac{\mathrm{d}i}{\mathrm{d}t} \quad . \tag{4.53}$$

The inductance therefore depends on the instantaneous quantum phase difference φ of the current i, according to the formula

$$L = \frac{\hbar}{2e}\,\frac{1}{I_c\cos\varphi} \quad . \tag{4.54}$$

This nonlinear reactance can be utilised for frequency conversion and for parametric amplification. The Manley-Rowe relations [4.9] normally indicate for nonlinear reactances relations between the powers which are transformed at various frequencies. Since the $\sin\varphi$ term in contrast to other loss-free reactances can also transform d.c. power into a.c. power, its power relations contain an additional term for the d.c. power. We assume that the $\sin\varphi$ term is embedded in a network which admits non-zero voltages only at frequencies f_1 and f_2 as well as at any frequency of the form $mf_1 + nf_2$ (m, n integers). The applied d.c. voltage U_{dc} corresponds according to (3.15) to the frequency f_{J} where

$$f_{\mathrm{J}} = kf_1 + lf_2 \tag{4.55}$$

(k, l integers). The power relations for the $\sin\varphi$ term are then [4.39], [4.40]:

$$\sum_{m=1}^{\infty}\sum_{n=-\infty}^{\infty} \frac{mP_{mn}}{mf_1 + nf_2} = -\frac{kP_0}{f_{\mathrm{J}}} \quad , \tag{4.56}$$

$$\sum_{m=-\infty}^{\infty}\sum_{n=1}^{\infty} \frac{nP_{mn}}{mf_1 + nf_2} = -\frac{lP_0}{f_{\mathrm{J}}} \quad . \tag{4.57}$$

P_{mn} is the real power flowing into the $\sin\varphi$ term at the frequency $mf_1 + nf_2$. P_0 is treated in the power relations just as the a.c. real power P_{kl} at the frequency f_{J}. These power relations show that it may be possible to apply Josephson junctions for d.c. pumped parametric amplification. Then, for example, in a circuit which is short-circuited for all alternating currents except at f_1 and f_2, it follows from (4.56) and (4.57) with $k = l = 1$ that

$$\frac{P_{10}}{f_1} = \frac{P_{01}}{f_2} = -\frac{P_0}{f_{\mathrm{J}}} \quad . \tag{4.58}$$

A d.c. power $|P_0|$ taken up by the $\sin\varphi$ term is delivered as a.c. powers $|P_{10}|$ and $|P_{01}|$ at f_1 and f_2. That this can be associated with a negative input resistance required for parametric amplification, we shall consider in the following way. For the case when the voltage in the $\sin\varphi$ term consists of a d.c. voltage and two superposed a.c. voltages, as

$$u(t) = U_{\mathrm{dc}} + \hat{U}_1 \cos(\omega_1 t + \varphi_1) + \hat{U}_2 \cos(\omega_2 t + \varphi_2) \tag{4.59}$$

the current is given, just as in (3.35), by

$$\begin{aligned} i_{\mathrm{J}}(t) = I_{\mathrm{c}} & \sum_{m=-\infty}^{\infty}\sum_{n=-\infty}^{\infty} J_m\left(\frac{2e\hat{U}_1}{\hbar\omega_1}\right) J_n\left(\frac{2e\hat{U}_2}{\hbar\omega_2}\right) \\ & \times \sin\left[\omega_{\mathrm{J}} t + \varphi_0 + m(\omega_1 t + \varphi_1) + n(\omega_2 t + \varphi_2)\right] \end{aligned} \quad . \tag{4.60}$$

For the phasors $\underline{I}_1, \underline{I}_2, \underline{U}_1$ and $\underline{U}_2$ of the currents and voltages at the frequencies f_1 and f_2 in $i_{\mathrm{J}}(t)$ and $u(t)$ one finds from (4.60) for small signals [4.40] that

$$\begin{bmatrix} \underline{I}_1 \\ \underline{I}_2^* \end{bmatrix} = \begin{bmatrix} 0 & Y_{12} \\ Y_{21}^* & 0 \end{bmatrix} \begin{bmatrix} \underline{U}_1 \\ \underline{U}_2^* \end{bmatrix} \tag{4.61}$$

with

$$Y_{12} = \mathrm{j}I_c \frac{e}{hf_2} \mathrm{e}^{\mathrm{j}\varphi_0} \ , \ Y_{21} = \mathrm{j}I_c \frac{e}{hf_1} \mathrm{e}^{\mathrm{j}\varphi_0} \quad . \tag{4.62}$$

From (4.61) and (4.62) it follows that at f_1 one measures the impedance

$$Z_1 = -\frac{h^2 f_1 f_2}{e^2 I_c^2} Y_2^* \tag{4.63}$$

looking into the sin φ term which is loaded by Y_2 at f_2. If Y_2 has a positive real part, then the real part of Z_1 is negative. The input reflection factor is then greater than one in magnitude, and an incident wave is reflected with amplification.

First attempts to realise low-noise parametric amplification by Josephson junctions were not very successful [4.41]. The large line width of the Josephson oscillation greatly raised the noise temperature. Only when the Josephson junction was connected in parallel with a circuit in series of low ohmic resistance R_R and an inductance L_R could low frequency noise currents be approximately short-circuited and hence the line width of the Josephson oscillation be significantly reduced [4.42]. Stable and low-noise amplification required that $L_R < 7\Phi_0/(2\pi I_c)$. With a signal frequency $f_1 = 10.35$ GHz the gain was $G > 5$ dB and the amplifier noise temperature $T_N < 25$ K.

Parametric amplifiers with Josephson junctions were investigated, not only with d.c. pumping, but also with radio frequency power pumping. With inductive parallel connection a gain of 16 dB was achieved according to [4.43] at $f_1 = 9.6$ GHz with $T_N = 18 \pm 35$ K, and according to [4.44] a gain of 10 dB at $f_1 = 8.2$ GHz with $T_N = 6$ K (measurement uncertainty of +15 K to −7 K) in a SQUID-based amplifier. Since a frequency conversion from the image frequency to the signal frequency can also occur with these parametric amplifiers, it should be pointed out that T_N is to be understood here as a single side band noise temperature.

Travelling wave amplifiers made from Josephson junction-type transmission lines have also been studied theoretically [4.45]. Here, as described by (3.35), a d.c. voltage and a static homogeneous magnetic field produced a Josephson phase-wave which worked as a pump-wave. There is parametric amplification if the sum of the signal and the auxiliary frequencies is equal to the pump frequency f_J and the sum of the wave numbers of the electromagnetic waves at f_1 and f_2 is equal to the wave number of the Josephson phase-wave.

From distributed Josephson junctions, amplifiers can also be constructed which work like field effect transistors [4.46,47]. They have an input port and an output port; a separation of forward and backward travelling waves as in amplification by a negative differential resistance is therefore unnecessary. This dual field effect transistor (FET) therefore appears particularly interesting.

Figure 4.11a illustrates the principle of the construction. By means of an external magnetic field H_0 a phase-wave $\varphi(z,t)$, see (3.53), is produced in the Josephson tunnel junction. H_0 can also be produced either wholly or partially by the input current I_S in the superconducting strip line of width w_S. According to (3.54) and (3.55) a d.c. voltage

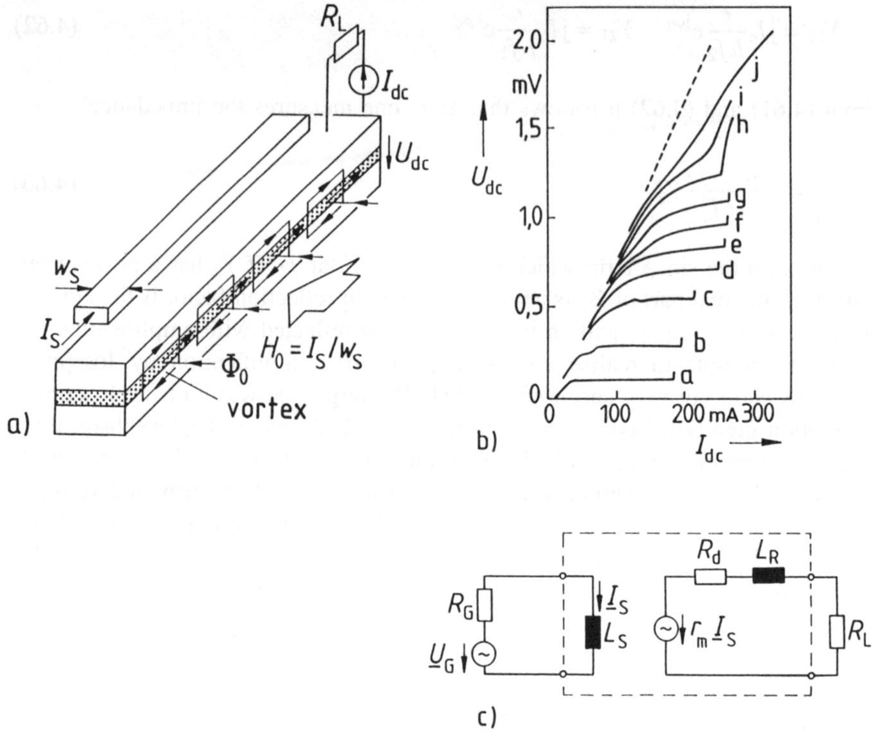

Fig. 4.11. Dual FET-amplifier, according to [4.47]. (**a**) Construction with moving vortices in phase wave; (**b**) current-voltage characteristics with external magnetic field H_0 as parameter. H_0 in A/m: a 178.4; b 201.6; c 235.2; d 268.8; e 302.4; f 336.0; g 396.6; h 436.8; i 504.0; j 604.8; (**c**) a.c. equivalent circuit with generator and load. L_S and L_R are inductances conditioned by the construction at the input and the output

$$U_{dc} = -\frac{\partial\varphi/\partial t}{\partial\varphi/\partial z} d' \mu_0 H_0 \tag{4.64}$$

between the electrodes is related to H_0. Here

$$-\frac{\partial\varphi/\partial t}{\partial\varphi/\partial z} = \frac{\omega_J}{k_z} = v_0 \tag{4.65}$$

is the velocity of the phase-wave. For small currents I_{dc} the wave number k_z is given by (3.55). For large currents I_{dc}, however, the equation $H_y =$ const. , from which (3.55) was derived, is no longer valid. Then v_0 tends instead to v from (3.41), i.e. to the propagation velocity of an electromagnetic wave in the Josephson junction. Under these conditions, with $H_0 = I_S/w_S$ it follows from (4.64) that

$$U_{dc} = \frac{v d' \mu_0}{w_S} I_S = r_m I_S \quad . \tag{4.66}$$

In order to guarantee this relation, I_{dc} must be really large, but the critical current I_c of the Josephson junction must not be exceeded. Figure 4.11b shows an experimentally produced family of curves. Here U_{dc} is plotted against I_{dc} with H_0 as parameter.

Figure 4.11c shows the a.c. equivalent circuit of the amplifier with the dual FET. The input impedance is purely reactive and is formed by the inductance of the control line with width w_S. The voltage $U_{dc} = r_m I_S$ with r_m from (4.66) is the no-load voltage of a controlled voltage source with internal resistance R_d. This is given by the gradient of the curve in Fig. 4.11b at the chosen operating point.

In the conventional FET a current density of electric charges e (As unit) is controlled by an electric field. In this dual FET the magnetic flux quanta $\Phi_0 = h/2e$ (Vs unit) transported by the phase-wave are controlled in their current density by a magnetic field. The FET is a voltage/current transformer with high input and output impedance. The dual FET is a current/voltage transformer with low input and output impedance. The current amplification

$$G_I = \frac{dI_c}{dI_S} = \frac{r_m}{R_d + R_L} \tag{4.67}$$

can, according to [4.48], amount to $G_I = 900\,\mu\text{m}/w_S$. In the dual FET, whose characteristics are shown in Fig. 4.11b, the limiting frequencies were estimated as 1 ... 5 GHz. This, however, must not be the upper limit in principle. Measured results relating to the noise are not yet available. Multi-stage amplifiers with dual FETs are discussed in [4.48].

As amplifiers for signal frequencies below the microwave region, amplifiers with DC-SQUIDs have also been investigated, see Sect. 3.4. Their input circuits are optimised for application as RF amplifiers for the operating frequency f_S. Thus according to [4.49] with $f_S = 93$ MHz and an operating temperature of 4.2 K a gain of 18.6 ± 0.5 dB and a noise temperature of only 1.7 ± 0.5 K were measured on a narrow band amplifier.

4.5 Oscillators

If a d.c. voltage $U_{dc} > 0$ is applied to a Josephson junction, an a.c. current of frequency f_J flows according to (3.15). Application as a voltage controlled oscillator is an obvious suggestion. Figure 4.12a shows the equivalent circuit diagram for a Josephson junction which obeys the RSJ model. For the load resistance $R_L = R_n$ is selected to maximise the output power. The voltage amplitude $\hat{U}_1$ at the frequency f_J is given by a Fourier analysis of (3.30), where R_n is replaced by $R_n/2$, since the effective parallel resistance in Fig. 4.12a is $R_n/2$. The power delivered at the load resistance at f_J is then $P_{L1} = \hat{U}_1^2/2R_n$, or $P_{L1} = K_1^2 I_c^2 R_n/2$ with $K_1 = \hat{U}_1/I_c R_n$. The behaviour of K_1 is plotted in Fig. 4.12b against the normalised d.c. voltage [4.50]. K_1 and hence P_{L1} increase monotonically with $\hat{U}_{dc}$ and hence with the frequency.

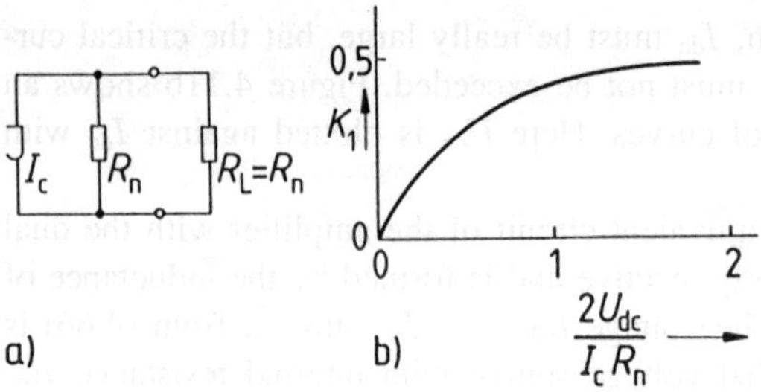

Fig. 4.12. Josephson junction in the RSJ model as oscillator. (a) Equivalent circuit; (b) Normalised RF voltage $K_1 = \hat{U}_1/I_c R_n$ versus the normalised d.c. voltage. From [4.50]

In order to find the order of magnitude of the power, we shall consider a typical example. With $I_c R_n = 1\,\text{mV}, I_c = 1\,\text{mA}$ and an operating point according to $U_{dc} = I_c R_n/2$ we find $R_n = R_L = 1\Omega, f_J = 242\,\text{GHz}, K_1 = 0.41$ and $P_{L1} = 8.4 \times 10^{-8}$ W. The available power is very small. The low normal resistance R_n moreover makes the match to the usual impedance level difficult. The line width of the oscillation without further precautions is according to (4.38) quite large. And the tunability is very limited because of the occurrence of harmonic oscillations [4.51].

A smaller line width and a higher output power are offered by the oscillator of [4.52–54]. It is constructed similarly to the dual FET-amplifier in Fig. 4.11a, but without the control line of width w_S. On the application of a d.c. voltage U_{dc} and a magnetic field H_0 magnetic vortices move in the direction of the load resistance R_L and produce an alternating current in it with frequency f_J, where according to (4.66) and (3.15)

$$f_J = \frac{2e}{h} v d' \mu_0 H_0 \quad . \tag{4.68}$$

According to (4.68) and the curves in Fig. 4.11b, with a constant direct current I_{dc}, the oscillator frequency f_J can be varied by the applied magnetic field H_0. According to [4.55] the line width of the oscillation is

$$\Delta f = \left(\frac{\pi k_B T}{\Phi_0}\right)^2 \frac{R_d^2}{R_0} \quad . \tag{4.69}$$

Here $R_d = dU_{dc}/dI_{dc}$ and $R_0 = U_{dc}/I_{dc}$ are the dynamic and the static resistance at the operating point in the characteristic of Fig. 4.11b. From (4.69) one finds typical line widths of only a few kHz at $f_J \simeq 10\,\text{GHz}$. The highest oscillatory frequency is expected at the energy gap frequency $4\Delta/h$ of the superconducting electrodes. With $f_J = 300\,\text{GHz}$ measurements give an output power of 4×10^{-6} W in a load resistance of $R_L = 1\,\Omega$. Such oscillators could find an application as local oscillators for SIS mixers.

4.6 Intrinsic Noise of Cryogenic Receiver Devices

As we have seen, superconducting components are suitable, amongst other things, for input circuits of microwave receiver devices with low intrinsic noise. In practice the development of instruments poses the question whether to use superconductor components or semiconductor components, which can if necessary be cooled to low temperatures. In this section both groups of components will accordingly be briefly reviewed to compare their noise characteristics.

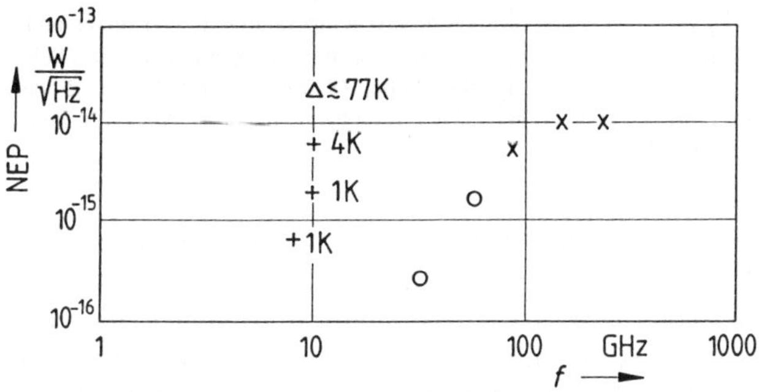

Fig. 4.13. Noise equivalent power NEP as a function of the frequency for microwave detectors at low temperatures. Δ Schottky, + super-Schottky-, o SIS-, × Josephson detectors

Figures 4.13 shows the experimentally achieved noise equivalent power NEP of quadratic microwave detectors as a function of frequency. As well as the SIS and broad-band Josephson detectors described above with operating temperatures $T \leq 4.2\,\mathrm{K}$, detectors with super-Schottky diodes are also described. These are Schottky diodes, hence rectifying metal–semiconductor junctions, whose metal electrode is a superconductor [4.65]. For traditional Schottky diodes operated as quadratic microwave detectors at low temperatures, few results are available. An indication is offered, however, by the value of $2.5 \times 10^{-14}\mathrm{W}/\sqrt{\mathrm{Hz}}$ measured at somewhat below 77 K and with $f = 10\,\mathrm{GHz}$ [4.57]. It may be added that traditional videodetectors, operating with traditional Schottky diodes at ambient temperature, show noise equivalent powers of about $5 \times 10^{-13}\,\mathrm{W}/\sqrt{\mathrm{Hz}}$ at 10 GHz and $2 \times 10^{-9}\,\mathrm{W}/\sqrt{\mathrm{Hz}}$ at 300 GHz. At the latter frequency the intrinsic noise of Josephson detectors is therefore about 5 orders of magnitude lower. Below 100 GHz SIS detectors and super-Schottky detectors give the least noise.

Figure 4.14 diplays the achievable equivalent single side band noise temperatures of receivers and mixers operated at low temperatures. The broken lines apply to mixer noise temperatures and the full lines to receiver noise temperatures. Nearest to the quantum limit of hf/k_B are the SIS mixers, for which results are so far available between about 30 and somewhat more than 300 GHz. The Josephson mixer has possible advantages at higher frequencies, the mixers with the super-Schottky diodes has advantages at lower frequencies. All three of

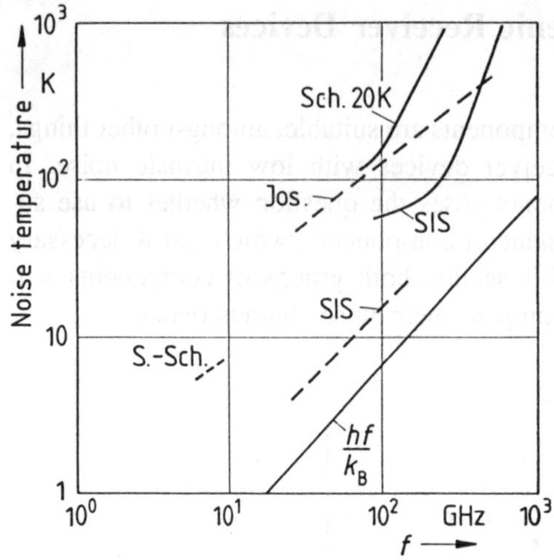

Fig. 4.14. Measured equivalent single side band noise temperatures of microwave mixers at low temperatures. (- - -) mixer and (—) receiver noise temperatures. (*Sch.*) Schottky diode, (*S-Sch.*) super-Schottky diode, (*Jos.*) Josephson junction, (*SIS*) SIS junction

these at present require operating temperatures with $T \leq 4.2\,\text{K}$. A mixer with a Schottky diode cooled to low temperature already achieves its lowest possible noise temperature at about $T = 20\,\text{K}$. It is significantly higher than that of the SIS mixer. Schottky mixers operating at room temperature give receiver noise temperatures between about 500 K at 30 GHz and 20000 K at 2 THz [4.58]. Compared with Schottky mixers, good SIS and Josephson mixers have smaller conversion losses, indeed, they can even show a positive conversion gain.

5. Materials and Production Methods

In this chapter we shall discuss various technological aspects of the production of SIS and Josephson junctions. Here we can discuss only the principles: for details one should refer to the relevant literature. We shall first in Sect. 5.1 broadly discuss materials and methods of production of tunnel junctions with superconducting electrodes, and the closely linked topic of planar superconducting circuits. Then in subsequent sections we shall deal with the special features of the production of microbridges and point contacts, and also of materials with higher critical temperatures.

5.1 Tunnel Junctions and Planar Superconducting Circuits

In this section we shall describe the choice of materials and methods of production for SIS sandwich structures and the associated planar superconducting circuits. The following applies to Josephson tunnel junctions and also in principle to SIS junctions. Since the latter, however, make use of a sharp knee in the quasiparticle branch of the current-voltage characteristic curves, the choice of electrode material for them has so far normally been limited to soft superconductors and their alloys.

Superconducting planar structures are produced by thin film technique, using the methods already well known as the basis of semiconductor technique, see [5.1]. In what follows we shall accordingly concentrate on the points which are peculiar to the production of superconducting circuits. Superconducting microwave circuits are built up from strip line or microstrip transmission lines. Here the substrate can either be used just as a base, or it may be used also as the dielectric of the transmission lines. This places different requirements on the relative permittivity and the dielectric loss factor of the substrate. Various substrates are compared in [5.2], and a few useful ones will now be briefly mentioned.

A substrate which is simple and also convenient for complicated circuits is glass [5.3] in the form of microscope slides. It has a smooth surface and is available in thicknesses of ≥ 0.1 mm. When the material parameters have to be closely defined, glass of a definite quality from a well-known manufacturer is sometimes preferred. Fused quartz and crystalline quartz are also used as substrates. In circuits with Josephson junctions silicon is very often found as the substrate. The reason for this is the great wealth of experience from semiconductor technique in the working of silicon. Many thin film technique instruments accordingly have their dimensions determined by the standardised sizes of silicon wafers. Because of its high thermal conductivity sapphire is also readily used as substrate material.

The methods usually employed today for the production of planar circuits with superconducting tunnel junctions are based to an important degree on the experience gained in the building of superconducting computers, see [5.4]. In what follows, however, we shall also indicate developments which were performed after those described in [5.4].

SIS sandwich structures are made by depositing two superconducting electrodes by a thin film technique, with an appropriate tunnelling barrier interposed between them. For the production of reliable tunnel junctions these barriers are quite critical. They can be produced as natural barriers by the oxidation of the base electrode, or as artificial barriers consisting of completely different material. If the barrier is formed of insulating material it must be, according to the particular application, about 1 to 5 nm thick. Since the normal resistance R_n, and in the case of Josephson tunnel junctions the critical current I_c also, depend exponentially on the barrier thickness, this barrier thickness must be defined with great precision. Moreover, in circuits with several similar Josephson junctions, it must be of the same thickness over the whole chip.

5.1.1 Electrode Materials

Table 5.1 from [5.5] lists the most important parameters of a few superconducting elements which may be considered for electrode material. They can be divided into two classes. First there are the soft metals such as Pb, Sn and In. These materials can be vaporised under a vacuum quite easily by ohmic heating. In the second class there are the transition metals such as Nb and V, of which niobium in particular is often employed. These materials, the hard superconductors, have high melting temperatures. To fabricate them as thin films is more expensive than for the soft superconductors.

The lifetime of a tunnel junction of hard materials is normally longer than that of a junction made of soft materials. For SIS junctions, however, mostly soft materials and their alloys have so far been used, since they normally have a more pronounced knee in the quasi-particle branch of the current-voltage characteristic

Table 5.1. Typical parameters for superconducting materials. The data relate to solid material. From [5.5]

Material	Critical temperature T_c [in K]	Critical mag. field H_c [in A/m]	Penetration depth λ_L [in nm]	Melting point in °C
Al	1.20	7.9×10^3	52	659
In	3.40	2.3×10^4	52	156
Sn	3.72	2.4×10^4	51	232
V	5.30	8.1×10^4	40	1920
Pb	7.19	6.4×10^3	40	328
Nb	9.26	1.6×10^4	44	2415

curve than the hard materials. Good elements using Nb electrodes and aluminium oxide as the barrier, however, are already showing a similarly well defined knee.

Of the soft superconductors, lead is readily used, since it can be vaporised at quite low temperatures, and at 4.2 K, the boiling point of helium at normal pressure, it clearly lies below its critical temperature. However, for regular use, i.e. outside the laboratory, pure lead does not qualify as electrode material, since the mechanical stresses occurring in repeated temperature cycling between room temperature and helium temperature (300 – 4 K) cause very severe damage to electrodes of pure soft material. Microscopically small hillocks and quasi-one-dimensional metallic whiskers form on the electrodes. These localised inhomogeneties can be much larger than the barrier thickness and make holes in the barrier. The tunnel junctions are then short-circuited and useless. Moreover the properties, especially the critical current I_c, of tunnel junctions with electrodes of pure lead change markedly within hours or days, if they are left exposed to air at ambient temperature. To a limited extent improvements can be achieved by, amongst other things, careful choice of substrate, protection by a sealing deposit of a layer of silica together with storage at low temperatures, e.g. at 77 K, the boiling point of liquid nitrogen, storage in a desiccator or in a vacuum.

A radical improvement, however, is achieved only if an appropriate additive is alloyed with the lead of the electrode. Such additives may have an effect not only on the internal structure of the electrode but also on the barrier, if it is produced by oxidation of the base electrode. The adhesion of the film to the substrate may also be heightened thereby. In [5.6] and [5.7] alloys with the following ingredients are recommended: base electrode Pb-In-Au with 12 % by weight of indium and 4 % by weight of gold, counter electrode Pb-Bi with 29 % by weight of bismuth. The latter electrode is dosed with bismuth not only to achieve good durability and invariance against temperature changes, but also because the quasi-particle residual current below the energy gap voltage can thereby be kept small. According to [5.8] this leakage current can be still further reduced if the gold in the base electrode is also replaced by bismuth. These elements have a very sharp knee in the quasi-particle tunnelling characteristic curve.

Further investigations show that the reliability of the tunnel junctions can be increased if the Pb-In-Au base electrode is deposited as a thin film on a thicker film of niobium. With an SiO protective layer five test chips, each with 11780 such tunnel junctions, showed no short-circuited junctions immediately after production. Even after 1000 temperature cycles between ambient temperature and 4.2 K none of the junctions was short-circuited. The scatter of the critical currents I_c on a chip was only trivially increased. Investigations on other samples [5.9], which were exposed in air over 400 days, showed that the critical currents had changed during this time by only from −10 % to +30 %.

Hard element superconductors can likewise be used with advantage as base electrodes and possibly also as the counter electrodes. They offer high mechanical strength, chemical stability, high critical temperatures, good adhesion to the substrate, high stability of the oxide together with no deformation of the electrodes even after many temperature cycles. A disadvantage of these materials is that they getter strongly during evaporation or sputter, i.e. that they absorb any residual gases present such as, for example, water vapour and oxygen before they can be precipitated on the substrate. Normally the critical temperature of the film is reduced thereby. The most frequently used of these materials in superconductor electronics is niobium. The production of tunnel junctions with both electrodes of niobium and niobium oxide as the barrier is described in [5.10]. Such junctions show a relatively large leakage current below the energy gap voltage [5.11]. The reason is presumably that between the oxide and the counter electrode a very thin superconducting layer (about 2 nm thick) is formed with a lower critical temperature [5.11]. Tunnel junctions with Pb-Bi as the counter electrode (29 % by weight bismuth) have only a quarter of this leakage current. About 16 000 of such junctions showed no drop-outs at all even after 14 000 temperature cycles over a few years [5.12]. When using niobium as the base electrode and a natural oxide (usually Nb_2O_5) as barrier, especially in high frequency applications, it should be remembered that the relative permittivity is very high ($\varepsilon_r \approx 29$) and therefore a capacitative shunt may easily occur.

Electrode materials made from alloys of hard superconductors are especially of interest for raising the critical temperature. A detailed review of the fundamental properties of tunnel junctions made of these materials is given in [5.13]. Of special interest is niobium nitride because of its high energy gap. For the making of tunnel junctions NbN is produced by reactive RF sputtering [5.14]. With both electrodes of NbN and an amorphous magnesium oxide film (MgO) as barrier, such tunnel junctions with a high energy gap voltage (5.1 mV) show an $I_c R_n$ product of 3.25 mV and a very small leakage current. A critical current is observed right up to temperatures of 14.5 K. The relative permittivity of MgO is $\varepsilon_r = 9.65$.

Table 5.2. Production Specifications for Superconducting Films, from [5.11]

Material	T_c in K	Crystal structure	Production method	Typical substrate	Substrate temp. in °C	Deposition rate in nm/s
Pb-In-Au	7	fcc	sequential therm.evap.	glass, silicon	−196...20	≃ 1.0
Pb-Bi	8.3	hcp	therm.evap.	counter electrode	0	≃ 3.0
Nb	9.3	bcc	sputtering, e-beam evap.	silicon	20 ...400	1... 2
Nb_3Sn	17.8	A15	e-beam evap.	sapphire	700 ... 800	1 ... 3
V_3Si	16.3	A15	e-beam evap.	sapphire	700 ... 800	1 ... 3
Nb_3Ge	23	A15	sputtering e-beam evap.	sapphire	750 ... 900	0.1 ... 1.5
Nb_3Al	16.7	A15	e- beam. evap.	sapphire	750 ... 1000	1 ... 9
NbN	17	B1	react. sputtering	sapphire	400 ... 600	2

Table 5.3. Material parameters for superconducting films, values in parentheses are estimates

Material	T_c in K	$\Delta(0)$ in meV	$\xi_{GL}(0)$ in nm	$\lambda(0)$ in nm	Ref.
Pb	7.2	2.7	510	32	[1.2]
Pb-In-Au	7	1.2	$\simeq 30$	150	[5.11]
Pb-Bi	8.3	1.7	$\simeq 20$	202	[5.11]
Nb	9.2	1.5	$\simeq 30$	85	[5.11]
Nb_3Sn	17.8	3.3	$\simeq 3$	170	[5.11]
V_3Si	16.3	2.5	$\simeq 3$	(150)	[5.11]
Nb_3Ge	23.6	3.9	$\simeq 3$	(150)	[5.11]
Nb_3Al	16.7	3.1	$\simeq 3$	(150)	[5.11]
NbN	$\simeq 16$	(2.4)	$\simeq 3$	(200)	[5.11]

Table 5.2, in most parts borrowed from [5.11], lists a few materials which are of interest for superconductor electronics. The fundamental conditions for the preparation of films from them are given. Table 5.3, also in most parts from [5.11], lists the relevant material parameters.

5.1.2 Tunnel Barriers

Oxide films are almost exclusively used as barriers for superconducting tunnel junctions. This offers a choice between a natural oxide, made by oxidation of the base electrode, or another metal oxide fabricated by thin film technique. The most important requirement for a reliable barrier is that the oxide is not porous. It then protects the base electrode against further oxidation.

On the other hand there are metals of which the oxide occupies a smaller volume than the material itself. This leads to tensile stresses and porosity. Then further oxygen molecules from the atmosphere are drawn into the metal, and the oxidation can continue. In the protective oxide the continuing oxidation can be described as a migration of positive metal ions through the oxide, which then are oxidised at the boundary layer with the atmosphere. A further oxidation process can occur at the metal-oxide interface by reaction of the oxygen ions which have passed through the oxide film. In both cases electrons diffuse from the metal to the oxidising atmosphere. These different processes are affected by the transport properties and the electrical properties, such as, for example, the diffusion constants of the metal ions and the oxygen ions, the electrical conductivity of the oxide and the charge distributions at the interfaces. For layers which are only some few nanometres thick the thickness of the layer in natural barriers increases logarithmically with time.

Thermal oxidation appears at first the simplest way of producing a barrier. The first electrode produced, the base electrode, is then exposed to an oxidising atmosphere such as air or pure oxygen. Various parameters can here play a part: oxygen pressure, oxidation time, substrate temperature, presence of water vapour and organic impurities, etc. If a clean surface is exposed to an oxidising atmosphere, the oxygen is first deposited either by chemical absorption as a monatomic layer or as stochiometric oxide on the surface. As the oxide film becomes thicker the distance between the oxide-oxygen interface and the oxide-metal interface increases. The movement of the ions through the oxide film is thereby reduced. The thermal oxidation process is therefore self-limiting. In this way uniform films of oxide can be deposited. Oxidation of the soft metals can take place at room temperature, oxidation of a niobium base electrode at 100 °C to 200 °C for a few minutes.

The thickness of the oxide barrier determines very sensitively the normal resistance and the critical current density of the superconducting tunnel junction. A longer oxidation time makes it easier to achieve accurately a stipulated critical current density. Longer oxidation times, however, also increase the probability that impurities are incorporated in the oxide layer.

Another oxidation method is the d.c. glow discharge or plasma discharge. In order to understand this, let us first notice that negatively charged oxygen ions are located at the oxide-oxygen interface. These assist the migration of the metal ions and promote the growth of the oxide film. With the growth of the thickness of the oxide film, however, the strength of the electric field decreases, since the two interfaces, oxygen-oxide and oxide-metal, are drawing further apart. The migration of the metal ions can now be assisted, however, by means of an externally applied electric field. For this the containing vessel is filled with an oxygen-containing gas at low pressure (about $10^2 \dots 10^4$ Pa). The glow or plasma discharge can be kindled by an electrode introduced with negative potential relative to its surroundings. The item to be oxidised, i.e. substrate with metal film, can either be mounted directly on the electrode or be placed elsewhere in the vessel. The latter, so-called "floating potential", method is widely used. Instead of a negative d.c. voltage, an RF voltage can be used for plasma discharge and oxidation.

In oxidation by plasma discharge gas ions can strike the target with such high kinetic energy that part of the target is eroded as if by sputter etching. *Greiner* [5.15] made use of this effect in the development of an oxidation method for lead base electrodes, in which the thickness of the oxide can be very closely controlled. It amounts to a method of RF plasma discharge, in which two mechanisms alternate in successive half-waves: oxide build-up and oxide erosion by sputter etching. The rate of the first process is roughly logarithmic, whereas the rate of erosion by sputter etching is approximately independent of the thickness of the oxide layer. By varying the oxygen pressure and the RF power the two processes can be separately controlled. In this way it is possible to achieve a state of dynamic equilibrium between the two competing processes, in which the thickness of the oxide layer is independent of time. This method can also be applied successfully to the oxidation of Pb-In-Au base electrodes. Figure 5.1 shows for this case, how the critical current density of Josephson tunnel junctions can, as in [5.7], be controlled by the oxygen pressure.

In the oxidation of niobium base electrodes a similar method can be used [5.10]. Unfortunately the growth of the oxide layer thereby is not self-limiting in the same pronounced manner as with base electrodes on a lead basis. For the fabrication of tunnel junctions by the oxidation of a niobium base electrode,

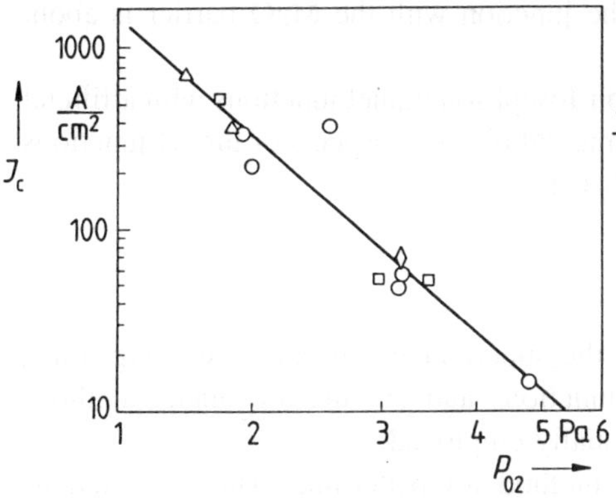

Fig. 5.1. Dependence of the critical current density J_c on the oxygen pressure in the Greiner process for Josephson tunnel junctions with Pb-In-Au base electrodes. Different symbols indicate various junctions which were deposited on base electrodes from different runs, from [5.7]

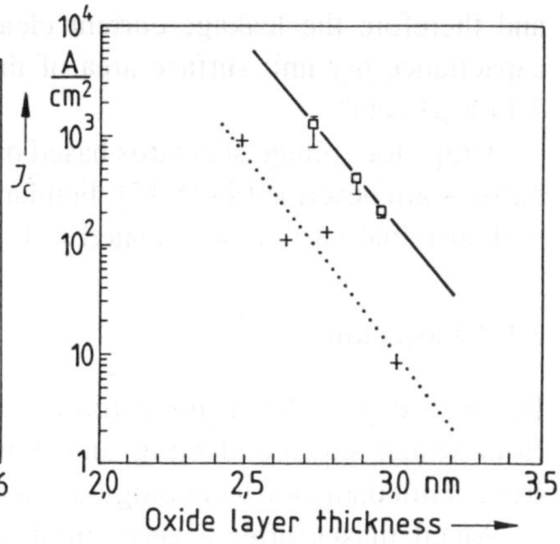

Fig. 5.2. Critical current density J_c as a function of the ellipsometrically determined thickness of the oxide layer on a Nb base electrode (□) cover electrode of Nb. (+) cover electrode of Pb, after [5.10]

therefore, the oxidation process has to be brought to an end after a definite time. Figure 5.2 shows how a further increase of the oxide layer by only an insignificant amount causes a drastic reduction in the critical current density.

Artificial barriers, i.e. those which are not produced by oxidation of the base electrode, offer the advantage of a greater choice of materials. Thus the leakage current in junctions with niobium electrodes can be drastically reduced by the use of aluminium oxide as the barrier. In [5.16] the fabrication process of such junctions is described in detail. Sputtering is used first of all to build up on the silicon substrate a film of niobium as the base electrode and then a film of aluminium a few nanometres thick is deposited. This aluminium film is then thermally oxidised at room temperature in an oxygen-argon atmosphere. Also sputtering is finally used to build up a film of niobium as counter electrode. Such junctions have an energy gap voltage of 2.8 mV and an I_cR_n product of 1.9 mV. With a current density $J_c = 500\,A/cm^2$ tunnel junctions of area 10 μm × 10 μm show a ratio of the quasi-particle tunnelling resistance below the energy gap voltage to the normal resistance of $R_{sg}/R_n = 46$. This high value shows how small the leakage current is. The critical current density of the junction can be controlled between 40 and 4600 A/cm² by the pressure and time of oxidation.

For the fabrication of Josephson tunnel junctions with electrodes of niobium nitride, magnesium oxide is a suitable artificial barrier [5.14]. First of all a NbN film is laid down on a silicon substrate by reactive RF sputtering. Then an MgO film is deposited on the NbN layer by RF sputtering from an MgO target. Finally the counter electrode is created by reactive RF sputtering from a Nb target in a nitrogen-argon atmosphere. Such junctions show an energy gap voltage of 5.1 mV and an I_cR_n product of 3.25 mV. The ratio $R_{sg}/R_n = 14$ is clearly larger and therefore the leakage current clearly smaller than with natural oxide. The capacitance per unit surface area of the junction with the MgO barrier is about 7 to 8 $\mu F/cm^2$.

Chips for voltage standards based on Josephson tunnel junctions with artificial barriers are described in [5.17]. Fundamental physical aspects of tunnel junctions with artificial barriers are treated in [5.13].

5.1.3 Patterning

In order to give the required form to the superconducting and insulating films, from which superconducting tunnel junctions and circuits are made, methods from semiconductor technology are usually employed.

Metal masks offer a very simple method for patterning. They are stencils made of thin sheet metal, through which the film is deposited. In order to obtain sharp edges the mask must be in close contact with the substrate. Good results are obtained with stainless steel masks, which are thin enough to be fabricated easily and to avoid any shadow effect, but strong enough to be mechanically stable. For fabrication of electrodes with lateral dimensions down to 0.2 mm a mask thickness of 0.05 mm is a good compromise. Such masks can be made by mechanical means or by a photo-etching technique. Electrodes as narrow as 1 μm, however, can also be deposited through metal masks [5.18]. Metal masks are less suitable for complicated designs.

Photolithography is the commonest method of fabrication for thin films. By this method lateral film measurements down to about 1 μm can be achieved. Here a photomask is used; this is a photographic plate with high resolution and high contrast. In contact exposure this photomask is laid directly on the substrate coated with photosensitive emulsion, which is called the photoresist. In projection exposure the design is projected through an optical system onto the photoresist.

The photomask can be made starting from a conveniently oversized drawing of the design, which is then transferred to the photographic plate via perhaps several photographic reduction processes. Instead of the drawing one often uses a red transparent plastic sheet, placed on a thicker transparent plastic sheet. The required design is cut out on a cutting board. This may be done either manually or under computer control. The superfluous portions of the red sheet are then removed. After careful photographic reduction the design is transferred to the photomask. A "step and repeat" camera can now replicate a design several times if necessary. In so-called pattern generators photomasks are directly exposed under computer control. The required design is fed into the computer using CAD techniques.

Figure 5.3 shows the most important steps in the photolithography. After a meticulous cleaning of the substrate it is completely coated under a vacuum with a superconducting layer. The substrate with metal film is then covered with a photoresist (1). By means of an emulsion centrifuge, the spinner, this film can be deposited very evenly and thinly (of order μm). Now it is exposed through the photomask using a mercury arc lamp. If, as in Fig. 5.3, a negative resist is

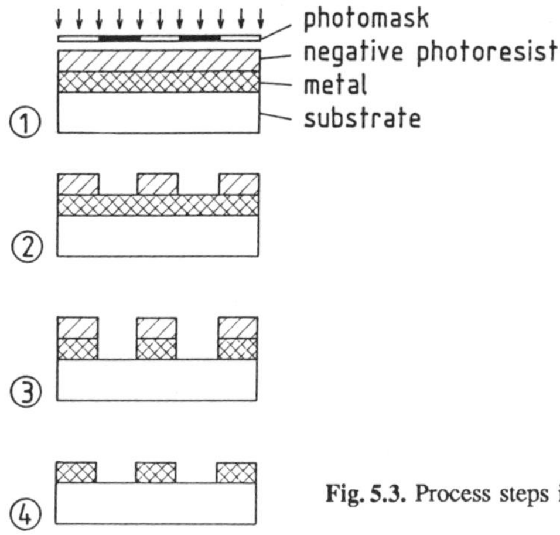

Fig. 5.3. Process steps in photolithography with negative photoresist

used, the unexposed regions are removed in the developing (2). For a specified thickness of the photoresist the optimal times of exposure and development have to be determined, in order to achieve the best possible sharpness of the edges. Where the superconductor is exposed it can then be etched away either chemically or by sputter etching. In the latter method the sample is used as the target, and the incident gas ions with high kinetic energy gradually remove the open parts of the superconducting film and parts of the photoresist. After the etching (3) the remaining photoresist is removed with an appropriate solvent (4). The same pattern of the film can be obtained by using a complementary mask and a positive photoresist. Then in the development process those regions are removed which had been previously exposed.

For the fabrication of some metal films the so-called lift-off process is convenient. As illustrated in Fig. 5.4, the substrate is first coated with a photoresist and then exposed through a photomask (1). After development the photoresist retains the pattern complementary to the required metal pattern (2). The metal layer is then deposited (3). Finally the photoresist beneath the metal film is dissolved away by a development process. The now unsupported regions of the metal film are thereby also lifted off, so that only those regions remain that are in contact with the substrate (4). In this last step it must be ensured that those regions of the metal which are to remain are not carried away with the rest. Accordingly the metal layer must be broken off at the edges. This can be achieved, for example, by treating the photoresist with chlorobenzol after the exposure. This causes the edges of the photoresist to become quite sharp or even contain a reverse taper. If in step (3) the metal hits the substrate perpendicularly the film thickness is considerably reduced at the edges because of the shadow effect. In the actual lift-off process the film then tears at this fracture point. Some further hints on the lift-off process are given in [5.19]. The lift-off method is in particular better adapted to the fabrication of many-layered thin film structures than the direct photolithographic method of Fig. 5.3.

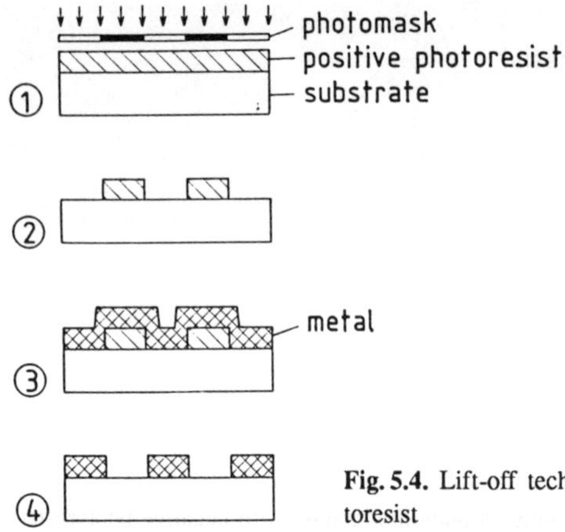

Fig. 5.4. Lift-off technique with photolithography and positive photoresist

In the fabrication of the counter electrode of superconducting tunnel junctions by the lift-off technique one is warned that microscopic short circuits to the base electrode can easily occur. Greater security against such microscopic short circuits is given by the window tunnel junctions of Fig. 4.2, where the tunnel surface is defined through an opening in a thicker insulating layer.

Another possibility for making the counter electrode without microscopic short-circuits, if it consists of niobium, is the SNAP process [5.20,21], see Fig. 5.5. First of all a large tunnel junction is fabricated over the whole silicon substrate without breaking the vacuum. Individual junctions are now excavated out of this sandwich by selective anodising of the niobium counter electrode. A mask of photoresist covers the parts which are not to be anodised (2). In the anodising the exposed niobium layer is oxidised in an electrolyte. After the selective anodising of the niobium a layer of SiO_2 to strengthen the isolator layer is deposited by RF sputtering and patterned by chemical etching (3). Finally a contact to the counter electrode is made, by depositing a further film of niobium by d.c. sputtering and patterning by plasma etching. This Selective Niobium Anodising Process SNAP can be used for a large number of material combinations of the base electrode and the barrier. Further developments of the SNAP process are the SNIP method for fabricating superconducting tunnel junctions with both electrodes of niobium nitride [5.22], and also the SNEP method for fabricating junctions with both electrodes of niobium, which however dispenses with the anodising process (SNIP: Self-aligned Niobium nitride Isolation Process; SNEP: Selective Niobium Etching Process).

Photolithography can be used to fabricate structures with lateral dimensions only down to about 1 μm, because if the dimensions are comparable with the wavelength of light diffraction effects occur and the contours are distorted. Electron beam lithography allows this limit to be lowered to about 0.1 μm. The exposure is then made with an electron beam. The associated wavelength is much smaller than that of visible light. Moreover electron beams can be deflected either electrically or magnetically. This offers the possibility of dispensing with

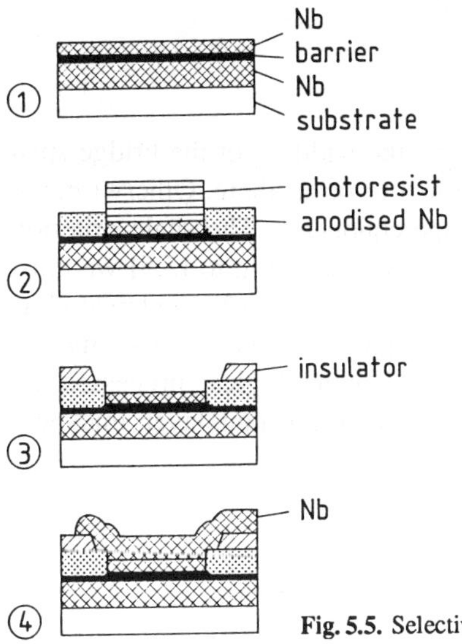

Fig. 5.5. Selective niobium anodising process SNAP, after [5.21]

photomasks and exposing the specimen directly under computer control. For the highest resolution one may also consider X-ray lithography.

In order to avoid the limitations apparently imposed by photolithography when fabricating the smallest superconducting tunnel junctions, one can also use rather special methods. An example of these is shown in Fig. 5.6. A layer of Nb/Nb_2O_5 on a SiO_2 coated silicon substrate is structured by ion etching to create a sloping edge. A tunnel barrier is now created on the sloping edge of the niobium, and in this example a layer of lead alloy is overlaid as the counter electrode. Even if the patterning of the tunnel junction in the direction perpendicular to the plane of the drawing Fig. 5.6 is performed by ordinary photolithography, because of the small dimension h the area of the tunnel junction is much smaller than it is possible to achieve by ordinary photolithography.

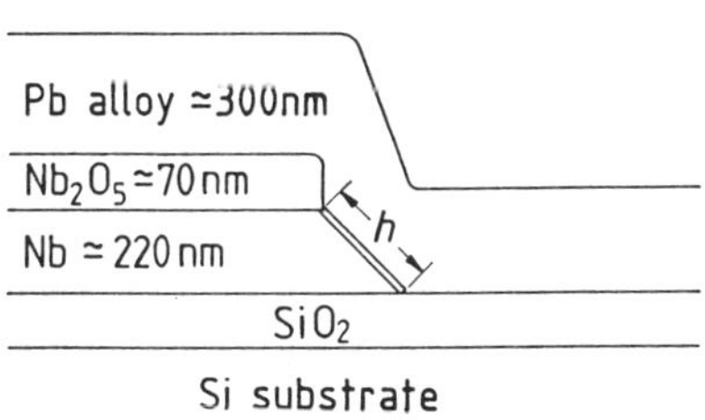

Fig. 5.6. Edge tunnel junction, after [5.11]

5.2 Microbridges

In a microbridge as in Fig. 5.7a the length L and the width w of the bridge span must be sufficiently small for the Josephson effect to occur there. Otherwise, the bridge span only behaves as a constricted superconductor. The scale criterion is therefore the Ginzburg-Landau coherence length. According to [5.5] we must have $L \simeq \max[L, w] < \xi_{GL}$, where the "less than" sign is to be understood as "approximately less than". Since ξ_{GL} is smaller, and mostly very much smaller, than $1\,\mu$m (see Tables 1.1 and 5.3), complicated technological processes are needed in order to fabricate such microbridges with constant layer thickness (Dayem bridges).

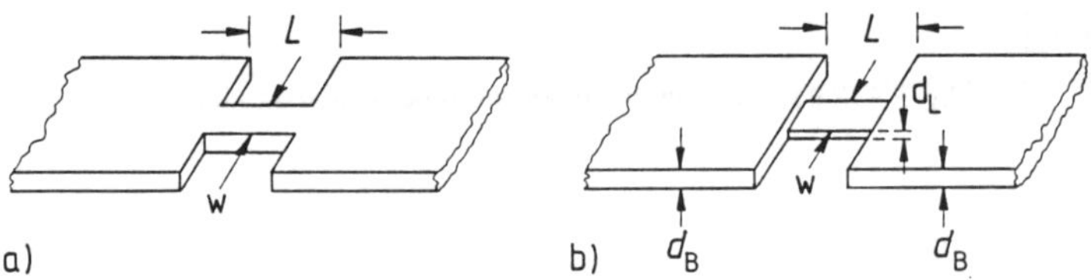

Fig. 5.7a,b. Microbridges. **(a)** Microbridge of constant thickness (Dayem bridge); **(b)** microbridge of variable thickness (variable thickness bridge, VTB)

In microbridges with variable layer thickness (**V**ariable **T**hickness **B**ridges, VTB), illustrated in Fig. 5.7b, the requirements of the geometry are reduced. For these we need $d_L \ll \min[d_B, l_B], \xi_{GL}, L$, in order that the Josephson effect should occur. Here d_L and d_B are the film thicknesses of the actual bridge span and of the abutment, respectively, and l_B the free path length in the abutment. The limitations regarding L and w are not so severe as for a bridge with constant thickness. For technological reasons the thickness of a film can be kept small more easily than its lateral dimensions.

The conditions for microbridges with Josephson effect can be most easily fulfilled by using pure soft superconductors with their relatively large values for ξ_{GL}. In a laboratory fabrication process, an appropriate metal film can be deposited on a substrate, and then the film can be eroded with a very fine needle so that the microbridge is left. This process, however, appears unsuitable for industrial production. One then has to consider other methods of fabrication, such as for example, electron beam lithography. Moreover the soft superconductors have the reliability problems already mentioned in Sect. 5.1.1.

Of greater technical interest are microbridges of soft superconductor alloys and of hard superconductors. For them, however, the coherence length ξ_{GL} is significantly smaller. Microbridges of these materials are usually made by electron beam lithography. But strictly speaking even with that the patterning possibilities are still too coarse.

Again, of the hard superconductors, the most interesting technically are those which have a high critical temperature, since operating with these requires comparatively little cooling. An exaple of these are the microbridges of Nb_3Ge in Fig. 5.7a. Their fabrication and their electrical behaviour are described in [5.24]. They are made from Nb_3Ge films, 30 to 150 nm thick, by electron beam lithography and reactive ion etching. The original films are superconducting up to 20 K. PMMA is used as the electron sensitive coating instead of the photoresist normally used in photolithography.

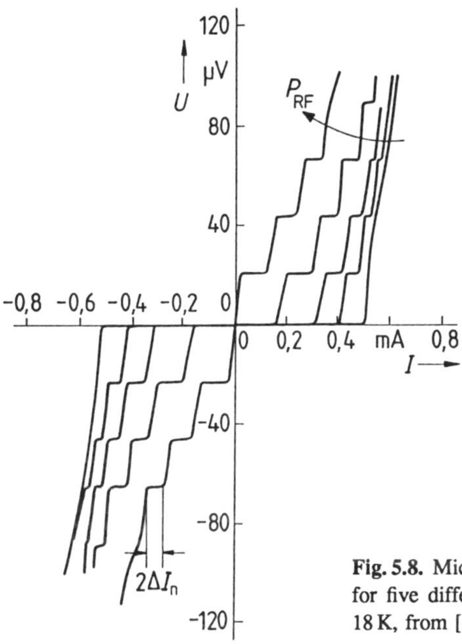

Fig. 5.8. Microwave-induced steps on a Nb_3Ge microbridge for five different microwave powers P_{RF}. l = 10 GHz, T = 18 K, from [5.24]

The bridges have widths w with 150 nm$< w <$ 300 nm and lengths $L \simeq$ 200 nm. Although both dimensions are clearly greater than the coherence length $\xi_{GL} \simeq$ 3 nm and therefore the above mentioned condition is not fulfilled, these microbridges, at least at reasonably high temperatures, show Josephson behaviour. According to Fig. 5.8 microwave-induced steps are observed, whose width, even with some periodicity, depends on the microwave power. The exact reason for this is not quite clear.

The requirement for extremely small dimensions of the bridge span can be somewhat reduced if the span itself consists not of superconducting but of normally conducting material. Then the bridge span becomes superconducting by the proximity effect. This effect allows superconductivity to diffuse a short distance into a contiguous normal conductor. The coherence length in a few pure normal conductors can amount to a few hundred nanometres. The critical current of such SNS bridges only decreases drastically if the length L of the normally conducting span between the superconducting abutments exceeds this coherence length.

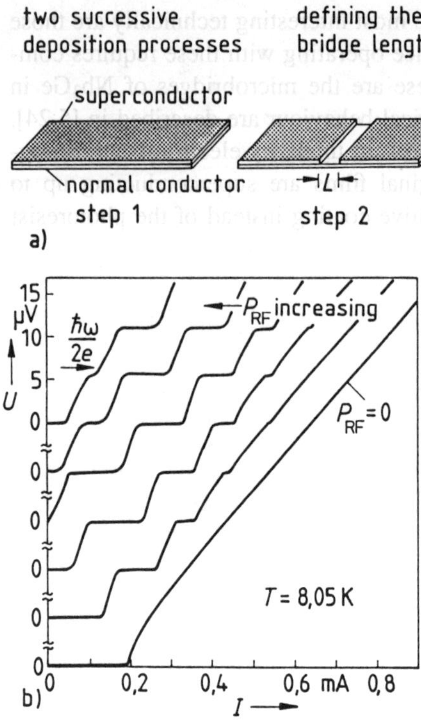

Fig. 5.9a,b. SNS bridge, from [5.25, 511]. (a) Fabrication method; (b) microwave-induced steps

An example of such an SNS bridge of Nb/Cu/Nb is shown in Fig. 5.9. According to [5.25] and [5.11] no lift-off process is used in their fabrication. They can therefore also be made with such superconducting materials as need a high substrate temperature for their deposition. A normal conductor-superconductor double layer is first fabricated as in Fig. 5.9a. In a first etching process a strip of the niobium is removed by projection photolithography and plasma etching. The length of the span is thereby set at somewhat less that 0.5 μm. In a second etching process the normal conductor layer forming the span is narrowed by photolithography and chemical etching. Figure 5.9b shows typical current-voltage characteristics with microwave-induced steps. $I_c R_n$ products up to 0.5 mV were achieved. SNS junctions have in general a very low resistance. In this case normal resistances were achieved up to $R_n \simeq 0.5\,\Omega$.

For further information on microbridges and methods for their fabrication reference should be made to [5.5] and [5.11], together with the references cited therein.

5.3 Point Contacts

The first investigations into the Josephson effect were carried out using point contacts. Even Josephson voltage standards were first implemented with point contacts.

The Josephson mixers described in the literature usually utilised point contacts. They have a smaller capacitance than tunnel junctions and are simpler to make than microbridges. Because of their three-dimensional character (see Fig. 3.5b) one can of course only with difficulty integrate them into planar circuits. Point contacts were therefore from the outset mechanically delicate and showed little reproducibility and stability in temperature cycling. One of the reasons for this was the imperfect matching of the expansion coefficients of superconducting materials and their mountings. The junctions therefore had to be adjusted in the cooled state. This meant that the pressure between the needle and the plate was set up manually by rods, levers and cogwheels from a screw at ambient temperature. This method is not very reliable.

In addition, the contact between needle and plate can be undefined. In the one limiting case there is a metallic conducting connection, in the other a tunnel junction with a thin oxide layer as barrier. In practice it is often a combination of both cases. Methods were therefore developed to fabricate point contacts which could be set up reliably and at room temperature before cooling [5.26]. Commercially operating SQUID magnetometers have for several years embodied such reliable point contacts. For microwave applications, however, these point contacts do not come up for consideration, since they have too low a normal resistance, a superconducting shunt or, with poured point contacts, an undefined dielectric environment. For application in microwave mixers, therefore, special designs of point contacts were developed. Two of them will now be explained.

Figure 5.10a shows a permanent point contact from [5.27] in a Sharpless wafer. This is a metal disk with a rectangular window which has the internal measurements of the rectangular waveguide, across the longitudinal axis of which it is installed. In the window is the point contact. The leads to it are led through holes in the disk. There are band rejection filters several $\lambda/4$ long in the d.c./intermediate frequency transmission line. The point contact is located in a waveguide of reduced height. A flexible whisker of niobium 0.025 mm thick is connected by point welding to a post 0.75 mm thick (Fig. 5.10b). The whisker is electrolytically sharpened to a point, so that the tip is less than 1 μm across. The "plate" of the point contact consists of a polished niobium stud of 0.75 mm diameter.

The electrolytic sharpening of the whisker can be carried out according to [5.28] in a mixture of 60 % by volume of HF (hydrofluoric acid) and 40 % by volume of H_2SO_4 (sulphuric acid), under about 10 V d.c. voltage (positive pole at the whisker). A carbon rod serves as the counter electrode. The etching process lasts about one second. In the same electrolyte, according to [5.28], the "plate" also can be electropolished by short voltage pulses.

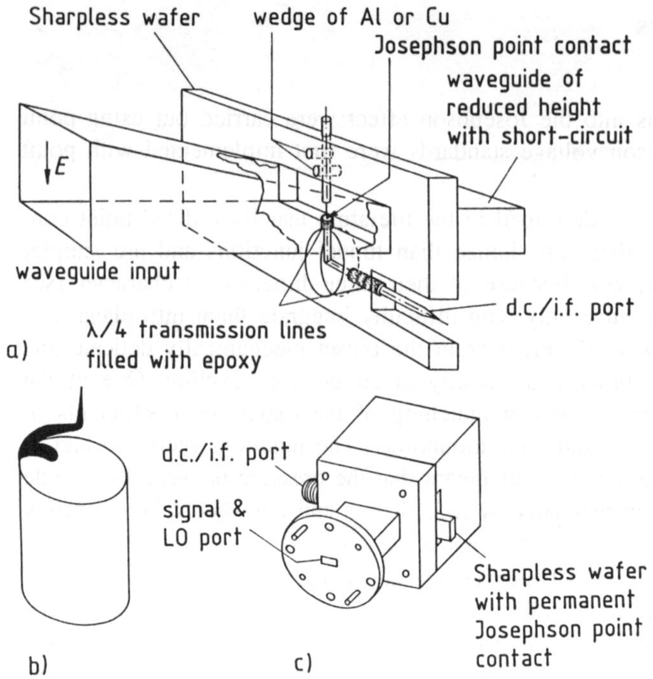

Fig. 5.10a-c. Permanent Josephson point contact in the Sharpless wafer from [5.27]. (**a**) Sketch of the principle. For impedance matching the waveguide height is tapered. The waveguide short-circuit lies about a quarter of a wavelength behind the point contact; (**b**) post with whisker 0.025 mm thick, sharpened electrolytically; (**c**) mixer block for 47 GHz

According to [5.27] the whisker is adjusted at room temperature with a special differential screw, to give a differential resistance of 30 to 70 Ω. Before the contact is set up, the electrodes are chemically cleaned with hydrofluoric acid in an atmosphere of helium gas. After cooling to 4 K this leads to a success rate of > 90 %. Typical values are $I_c = 28\,\mu$A with $R_n I_c = 0.92$ mV. The contact area was estimated at 0.1 μm^2. With 47 GHz injection microwave-induced steps up to $n \geq 12$ could be observed. Repeated temperature cycling between 300 and 4 K did not impair the behaviour of this permanent Josephson point contact.

The second design of a permanent Josephson point contact described here is shown in Fig. 5.11. According to [5.29] this contact is made of Nb wire and a Nb foil, which are mounted on two metallised substrates. These are stuck on a third substrate, see Fig. 5.11. For matching to the thermal expansion coefficient of niobium, the substrate material used was Corning-8260 glass. Planar metal structures for the band pass filters were deposited by photolithographic methods on the substrate. Then an angle of Nb sheet 12 μm thick and a Nb wire of 18 μm diameter were each attached by point welds at the ends of the metallised substrates. The Nb wire was electrochemically sharpened to a point radius < 0.25 μm. In assembling the junction the substrate with the Nb angle is first glued

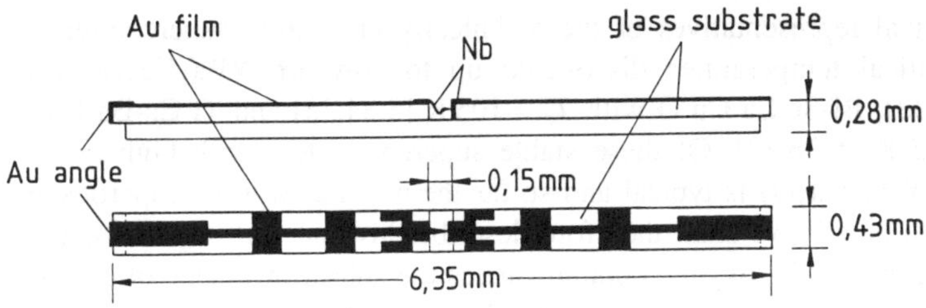

Fig. 5.11. Permanent Josephson point contact from [5.29]

to the base plate (glue: Eastman 910). Then the substrate with the Nb wire can be moved forward by a differential micrometer screw until a resistance of 20 to 30 Ω is observed. This substrate is then also glued to the base plate. For application as mixer the arrangement is installed through openings in the sides of a rectangular waveguide.

On cooling to 4.2 K the resistance usually increases by about 15 %. This change, however, is reversible and reproducible. The current-voltage characteristic of such a Josephson point contact was checked over more than ten temperature cycles for a month. It showed no change. At $T = 1.8\,K$ the $R_n I_c$ product amounted to 0.6 to 0.8 mV.

For further information on the fabrication of Josephson point contacts one should consult [5.30] and [5.5].

5.4 Oxide Superconductors with High Critical Temperatures

After H. K. Onnes had discovered superconductivity in 1911 in experiments with mercury ($T_c = 4.15\,K$), materials were gradually found and developed with higher critical temperatures. For more than a decade, however, a limit appeared to have been reached with $T_c = 23.2\,K$ of Nb_3Ge. Systematic investigations of metallic oxides by Bednorz and Müller brought in 1986 the first indication that critical temperatures around 30 K could be achieved with chemical compounds of Ba-La-Cu-O [5.31]. On this foundation a large number of further studies was undertaken, also using different materials. Although by now (April 1989) material combinations with critical temperatures up to above room temperature, namely up to 338 K (65 °C), have already been reported [5.32], it appears that till now only superconductors with critical temperatures up to about 125 K are sufficiently stable and reproducible to be of technical interest. Nevertheless, the critical temperatures of these superconductors lie above the boiling temperature of 77 K for liquid nitrogen. Their discovery has therefore quite significantly reduced the expense of refrigeration needed to achieve superconduction - see also Chap. 6. This important development was made possible only by the fundamental investigations of Bednorz and Müller. They were accordingly awarded the 1987 Nobel Prize for Physics.

Three typical representatives of the technically interesting superconductors with high critical temperatures discovered up to now are $YBa_2Cu_3O_7$ with $T_c = 92\,K$ [5.33], Bi-Sr-Ca-Cu-O with $T_c = 105\,K$ [5.34,35] and Tl-Ca-Ba-Cu-O with $T_c = 125\,K$ [5.36,37]. Of these stable superconductors with high critical temperatures $YBa_2Cu_3O_7$ is typical and so far the material whose properties are known in most detail. We shall therefore describe this material in particular. In the chemical formula the oxygen content carries the subscript 7. One often finds the subscript 7-δ, as it differs slightly from 7 in the optimal composition.

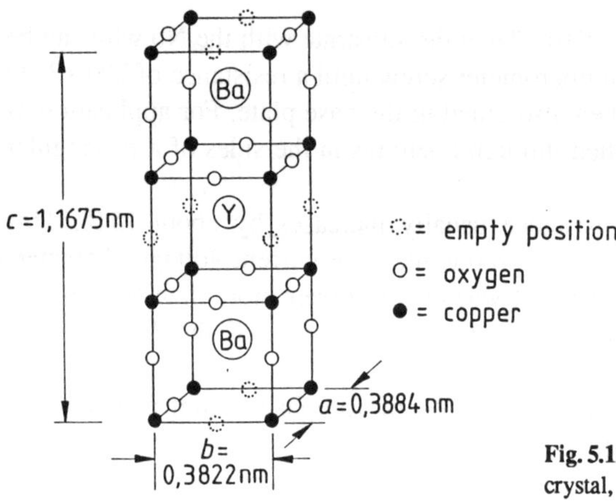

Fig. 5.12. Lattice structure of $YBa_2Cu_3O_7$ crystal, from [5.38]

In $YBa_2Cu_3O_7$ the crystal structure is evidently quite critical for its superconduction properties. Figure 5.12 shows this crystal structure [5.38]. It is quite similar to a perovskite structure. Only a few lattice positions are unoccupied, which would be occupied by oxygen atoms in the complete perovskite structure. In Fig. 5.12 these empty places are shown dotted.

In the $YBa_2Cu_3O_7$ crystal, planes of copper oxide lie on top of one another. The stratified structure also leads to a strong anisotropy in properties. According

Table 5.4. Anisotropy parameters for a $YBa_2Cu_3O_7$ crystal at $T = 11\,K$, from [5.39,40]

$\xi_{GL,ab}/nm$	2.2	
$\xi_{GL,c}/nm$	0.67	
	$H \parallel c$	$H \parallel a,b$
$\lambda_{a/b}/nm$	95	780
λ_c/nm		95
κ	44	230
$\mu_0 H_{c1}/mT$	69 ± 5	12 ± 1
$\mu_0 H_{c2}/T$	70	230
$J_{c,ab}/(Acm^{-2})$	6×10^7	1×10^7
$J_{c,c}/(Acm^{-2})$		1.2×10^6

to Table 5.4, which is taken from [5.39,40], the Ginzburg-Landau coherence length ξ_{GL} in the a, b plane is distinctly larger than in the c direction of the crystal. In both cases, however, they are only of the order of a few atom separations. The superconduction penetration depth λ is also very different in the a, b plane and in the c direction, and moreover depends upon whether the magnetic field is in the direction of the c axis or in the a, b plane. For both field directions the Ginzburg-Landau parameter $\kappa = \lambda/\xi_{GL}$ is very much larger than unity. It is therefore decidedly a superconductor of the second type, see Sect. 1.6. The lower critical magnetic field H_{c1} for $H \parallel c$ and $H \parallel a, b$ lies at $69\,mT/\mu_0$ and $12\,mT/\mu_0$, respectively, and the upper critical magnetic field H_{c2} at the extremely large values of $70\,T/\mu_0$ and $230\,T/\mu_0$, respectively. The critical screening current densities J_c corresponding to these values vary according to the current direction and the magnetic field direction by more than an order of magnitude.

It should be pointed out here that the thermodynamically critical magnetic field H_c of $YBa_2Cu_3O_7$ for $T = 87\,K$ amounts to about $103\,mT/\mu_0$ and, assuming a BCS type profile, increases at lower temperatures to about $1\,T/\mu_0$ at $T = 0\,K$ [5.41].

Measurements of the energy gap of $YBa_2Cu_3O_7$ give results which are scattered by more than a factor of 2 [5.42]. They vary about a value of $2\Delta(0) = 30\,meV$. This value was determined from tunnel measurements and corresponds to (1.44) with $T_c = 92\,K$ and the factor 3.9 instead of 3.52 [5.43]. Since (1.44) is quite well satisfied by this and since (1.44) stems directly from BCS theory, the superconduction in $YBa_2Cu_3O_7$ will presumably also be well described at least by parts of BCS theory. Also the measured temperature dependence of the energy band gap $\Delta(T)$ appears to agree with that following from BCS theory, see (1.43) and Fig. 1.12 [5.43].

The material properties measured on a single crystal of $YBa_2Cu_3O_7$ in general are not reached by those of technical materials, such as bulk material, wires, layers and thin films. The method of manufacture therefore plays quite an important role.

Bulk material can be manufactured by a chemical solid state reaction. According to [5.44] a cold-pressed mixture of $Y_2O_3, BaCO_3$ and CuO is annealed for 12 hours at 950 °C in air. Then follows a concluding heat treatment for 12 hours at 700° to 800 °C in an atmosphere of oxygen. A similar procedure, involving repeated annealing, cooling and pulverisation, is given in [5.45].

The specific resistance ϱ, measured on small rods ($1 \times 1 \times 7\,mm^3$) of ceramic manufactured according to [5.44], is shown in Fig. 5.13a. $\varrho = 0$ occurs for this specimen at $T = 91.0\,K$. The width of the transition from the metallic to the superconducting state (fall from 90 % to 10 % of the measured signal) is 1.2 K. Other specimens showed $\varrho = 0$ up to $T = 95\,K$.

The measured a.c. susceptibility χ_{mac}, determined from inductivity measurements with magnetic flux densities of about $1\,\mu T$, is plotted in Fig. 5.13b. The transition width from the normally conducting to the superconducting state (again 90 % to 10 %) amounts only to 0.45 K.

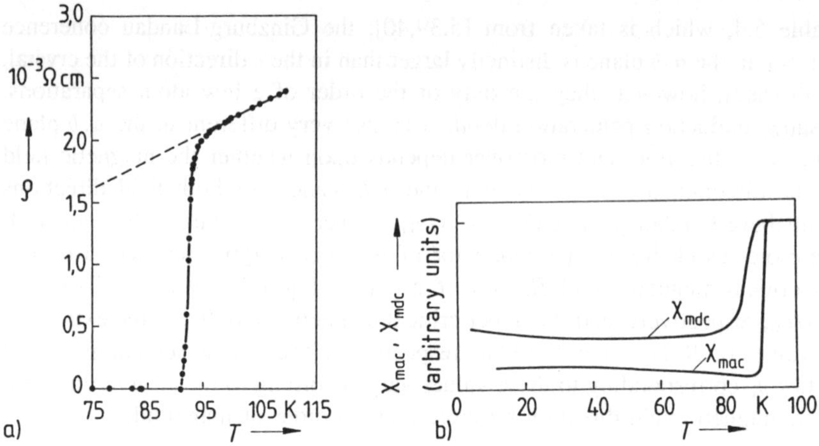

Fig. 5.13a,b. Measured temperature dependence of the properties of bulk $YBa_2Cu_3O_7$ which was made in a solid state reaction, after [5.44]. (**a**) specific resistance ϱ, (**b**) magnetic a.c. and d.c. susceptibility χ_{mac} and χ_{mdc}, respectively

The flux exclusion arising from the Meissner effect was determined using a SQUID magnetometer with magnetic d.c. fields of 2 mT. For this the specimen was placed in the detector coil of the SQUID and cooled, starting at above T_c. The best specimen gave, between 3.5 and 80 K, a diamagnetic signal which corresponded to 25 % to 40 % of complete diamagnetisation. This signal is also plotted in Fig. 5.13b as the d.c. susceptibility χ_{mdc}. The transition width is greater than for the $\varrho(T)$ and the $\chi_{mac}(T)$ profiles. This and the incomplete diamagnetisation are explained by the fact that the bulk $YBa_2Cu_3O_7$ made in the solid state reaction is a granular superconductor. It consists of grains whose interior is superconducting, but whose surfaces are coated with a thin layer whose composition is different from that of the interior [5.45]. This layer is normally conducting or semiconducting, or even insulating. Most are normally conducting [5.46]. Accordingly, where the grains touch, Josephson junctions are formed with the layers as barriers. The granular superconductor consists of a three-dimensional network of many Josephson junctions connected together. Compared to a homogeneous superconductor, significant differences in the physical properties can occur.

For superconducting electronic circuits, still more interesting than the bulk material are thin films. They can be made, for example, by sputtering [5.47,48], evaporation [5.49,50] and molecular beam epitaxy [5.51]. For a literature survey see [5.52]. The quality of the film is primarily determined by the critical temperature for vanishing electrical resistance, the magnetic susceptibility and the critical current density. The highest possible critical current densities are of primary interest for power and high magnetic field applications of the superconductors. However, also for the realisation of superconducting electronic components, certain minimum values differing from case to case are required. For microwave applications the surface resistance R_S of the film is of special interest.

The original method of manufacture of thin films from superconductors with high critical temperatures consists of depositing on a substrate the components, e.g. Y, Ba and Cu as an amorphous mixture of metals and oxides in the presence of oxygen. This amorphous mixture is annealed at high temperatures ($\simeq 900\,°C$), whereby the required superconducting phase is formed. Finally the film is cooled slowly in an oxygen atmosphere, thereby incorporating the correct content of oxygen. If this procedure is applied with a suitable substrate, a strongly oriented film is formed in an epitactic growth process. With a substrate of a single crystal of strontium titanate $SrTiO_3$ with the [100]-direction perpendicular to the deposition plane of the film, most high T_c crystallites align themselves with their c-axis perpendicular to the film plane [5.49].

In general these films are the more high grade the better they are oriented and the fewer grain boundaries there are. In both respects one can achieve improvements if one makes the film *in situ*, hence avoiding a separate annealing process. The substrate must then indeed be heated during the deposition process, but not to so high temperatures as with separate annealing. Such *in situ* processes can be based, according to [5.53], on electron beam or thermal evaporation [5.54–58], sputtering from a target of suitable composition [5.59,60], ion beam sputtering or laser evaporation from a target of suitable composition [5.61,62]. In a few cases the presence of molecular or activated oxygen is necessary.

As an example the method using laser evaporation according to [5.62] will now be described. An XeCl excimer laser with a wavelength of 308 nm, 60 ns pulse duration, a pulse rate of 5 Hz and a pulse energy of 2 J is used. The laser beam is focused through a window on to a rotating $YBa_2Cu_3O_7$ sinter target. The vaporised material is deposited on to a $SrTiO_3$ substrate about 30 mm distant, heated to about 750 °C. After the deposition of the film the substrate is gradually cooled in an oxygen atmosphere. A critical temperature of 90 K is achieved and at $T = 77$ K a critical current density in the film of $2.2 \times 10^6\ A/cm^2$. The critical current density of amorphous films or indeed of bulk sintered material lies orders of magnitude below this value.

Films made according to [5.62] had their surface resistance R_S measured at 86.7 GHz [5.63]. The better of two specimens showed a surface resistance $R_S < 8\ m\Omega$ at $T = 77$ K. According to Fig. 5.14 it therefore lies at least an order of magnitude below that of copper at the same temperature and frequency. Polycrystalline material has a much greater surface resistance.

As well as the d.c., the a.c. Josephson effect was also observed with elements of $YBa_2Cu_3O_7$. Of special interest here is the behaviour at temperatures $T \geq 77$ K. In order not to be greatly suppressed by thermal noise, the critical currents of the Josephson junctions must be greater than about 65 μA. Then one gets the theoretical dashed line shown in Fig. 3.6 for $\gamma = 40$ with rounding of the current-voltage characteristic. This case can be regarded as the limit for technical applicability.

The Josephson effects were observed on bulk $YBa_2Cu_3O_7$ and on films of this material. Here the geometry of the superconductor need not even be constricted. Then because of the granular structure there are many Josephson junctions present

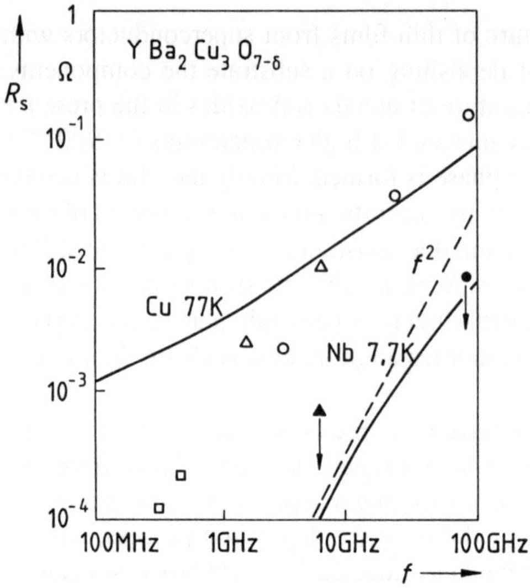

Fig. 5.14. Frequency dependence of the surface resistance R_S at 77 K for polycrystalline (*open symbols*) and single crystal (*solid symbols*) $YBa_2Cu_3O_7$ measured at Argonne (*squares*), Cornell (*triangles*), Wuppertal (*circles*). Curves for copper and niobium are shown for comparison. From [5.63]

in the bulk material. However, constrictions make the Josephson effects stand out more clearly. Thus, at T = 77 K microwave induced steps (f = 35.5 GHz) were observed on a bridge of $YBa_2Cu_3O_7$ which, by sawing a bulk block, was left standing as an island with a cross-sectional area of 1 mm^2 and a length of 0.5 mm [5.64]. The maximal R_nI_c at 0.2 mV lay far below the energy gap voltage of 30 mV. Also with films which are constricted in the microbridge manner, microwave induced steps can be observed. The dimensions of the bridge may be about three or four orders of magnitude greater than the coherence length ξ_{GL}. The presence of a few grain boundaries with SNS character then makes Josephson effects possible. For example, in [5.65] granular bridges about 10 μm × 10 μm were prepared using the lift-off technique from $YBa_2Cu_3O_7$ films made by sputtering. The grains were oriented with respect to their c-axes and had typical lateral dimensions of 4 μm. At frequencies of 9.37 and 55 GHz sharp steps of constant d.c. voltage could be observed in the U(I) characteristic up to temperatures of 71 K.

An RF-SQUID, functioning up to a temperature of T = 81 K, is described in [5.66]. There a bulk ring of $YBa_2Cu_3O_7$ ceramic was broken at a breakpoint and then pressed together again, so that one or more Josephson junctions are formed here. The magnetic flux through the ring can then be measured, see Sect. 3.4. At T = 75 K there was the intrinsic characteristic noise of the RF-SQUID, which corresponds to a flux noise with a spectral density of only $4.5 \times 10^{-4} \Phi_0/\sqrt{\text{Hz}}$. Such a SQUID made by breaks, however, appears to have little reproducibility and stability.

In a technically more interesting variant, a cylindrical sintered body of $YBa_2Cu_3O_7$ was shaped, using drills and saws, so that a three-dimensional bridge with dimensions in the region of 0.1 mm was obtained. RF-SQUIDs constructed from this show a flux noise at T = 77 K down to $5 \times 10^{-4} \Phi_0/\sqrt{\text{Hz}}$ [5.67]. However, the $1/f$ noise increases strongly below about 100 Hz [5.68]. For a comparison of high T_c SQUIDs with each other and with commercial 4.2 K-SQUIDs, see [5.69]. This comparison also includes high T_c DC-SQUIDs.

The first DC-SQUID of $YBa_2Cu_3O_7$ in a thin film technique, see Sect. 3.4, was described in [5.70]. The superconducting film was there patterned by converting parts of the film to insulators by ion implantation of oxygen or arsenic. With a width and length each of about 17 μm, the dimensions of the resulting microbridge were about four orders of magnitude larger than the coherence length. Nevertheless this SQUID showed a periodic dependence of the maximal current on the magnetic flux up to temperatures of $T = 68$ K. The granular structure in the region of the bridge was presumably the reason for this. The flux noise at 40 K was estimated at $< 5 \times 10^{-5} \Phi_0 / \sqrt{\text{Hz}}$. According to [5.71] this value should also be achievable with an optimally adjusted DC-SQUID at 77 K.

The characteristic data on granular bridges of high T_c material will no doubt always show a high degree of scatter. Moreover, till now they show only small $R_n I_c$ products. A remedy could be found in the proposed manufacture of SNS bridges, as in Fig. 5.9, whose edges consist of high T_c superconductors. For the first bridge of this kind microwave induced steps of constant voltage have already been observed [5.72].

The manufacture of SIS sandwich structures with high T_c material has so far mostly failed. The reasons for this are possibly the small coherence lengths in the electrodes and the high process temperatures necessary in setting up the counter electrode; at this stage it is easy to destroy the previously manufactured extremely thin insulating tunnel layer. According to [5.73], however, success has already been achieved in the manufacture of a tunnel barrier by fluorine treatment of the $YBa_2Cu_3O_7$ base electrode, so that after making the $YBa_2Cu_3O_7$ counter electrode, single electrode tunnel curves with knees similar to those in Fig. 2.3 could be observed up to temperatures of 70 K. A Josephson current at $U = 0$ V, however, has not been recorded.

6. Low Temperature Technology

In this chapter we shall explain some fundamental aspects of low temperature technology. First of all we shall describe some methods for achieving low temperatures. Then we shall give some advice on how to deal with low temperatures, and finally a few examples of microwave systems which are operated at low temperatures. More details on low temperature technology are to be found in [6.1–5].

6.1 Generation of Low Temperatures

In this section we shall explain the fundamental principles of the most important methods by which one can obtain temperatures down to 2 K. To refresh the reader's memory, however, we shall first recall a few thermodynamic relations.

The cooling of a system to a temperature T_1, lying below the ambient temperature T_2, and the maintenance of this temperature in the system, requires the transfer of a quantity of heat Q_1 from temperature level T_1 to temperature level T_2. According to the second law of thermodynamics this transfer cannot take place spontaneously, but becomes possible only when work is performed by another system.

According to the first law of thermodynamics the energy delivered to a system as a quantity of heat $\mathrm{d}Q$ is used in the general case in the raising of its internal energy by $\mathrm{d}u$ and in the performance of external work $p\,\mathrm{d}v$ by that system, where

$$\mathrm{d}Q = \mathrm{d}u + p\,\mathrm{d}v \quad . \tag{6.1}$$

In an adiabatic process $\mathrm{d}Q = 0$ and hence

$$\mathrm{d}u = -p\,\mathrm{d}v \quad , \tag{6.2}$$

i.e. the expansion of any gas whilst performing external work leads to the lowering of the internal energy and hence to the lowering of the temperature of the gas.

In the mathematical treatment of cooling processes the concept of enthalpy arises. The enthalpy h can be obtained from the following relation:

$$\mathrm{d}h = \mathrm{d}u + \mathrm{d}(pv) = \mathrm{d}u + p\,\mathrm{d}v + v\,\mathrm{d}p \quad . \tag{6.3}$$

Using this equation one can put the first law of thermodynamics (6.1) in the following form:

$$dQ = dh - v\,dp \quad . \tag{6.4}$$

The second law of thermodynamics leads to the definition of the entropy s. In differential form this definition reads

$$ds = \frac{dQ}{T} \quad . \tag{6.5}$$

In an adiabatic process we have $dQ = 0$ and consequently s = constant, i.e. the entropy remains unchanged: the adiabats coincide with the curves of constant entropy.

If one wishes to cool an object continuously, removing a stipulated quantity of heat per unit of time, this heat has to be transferred from a low temperature level to a higher level. The cooling medium has to be returned to its initial state, in order to repeat the process. This cooling cycle consists of alternating elementary processes in which the cooling medium changes its state. Convenient cooling media are (in the order of cooling temperatures) ammonia, carbon dioxide, Freon, ethylene, methane, oxygen, nitrogen, neon, hydrogen and helium.

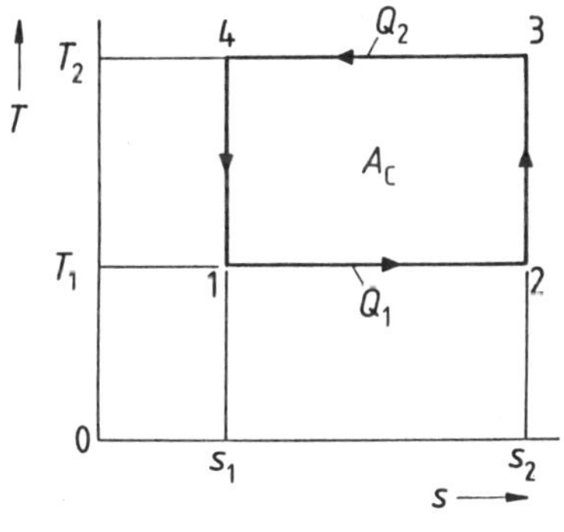

Fig. 6.1. Carnot cycle in the T, s diagram

The efficiency of actual cycling processes is measured against the reversible Carnot cycle. This consists (Fig. 6.1) of two isotherms and two adiabats. The quantity of heat Q_1 is extracted from its surroundings at constant low temperature T_1 by isothermal expansion along the line 1,2. The cooling medium is then adiabatically compressed (line 2,3), which raises its temperature to the value T_2. At this temperature the quantity of heat Q_2 is lost to the surrounding environment by isothermal compression along the line 3,4. Q_2 is calculated from the equation

$$Q_2 = Q_1 + A_C \quad . \tag{6.6}$$

Here A_C denotes the work performed on the system in the cyclic process. The cooling medium then expands adiabatically (line 4,1), whereby its temperature

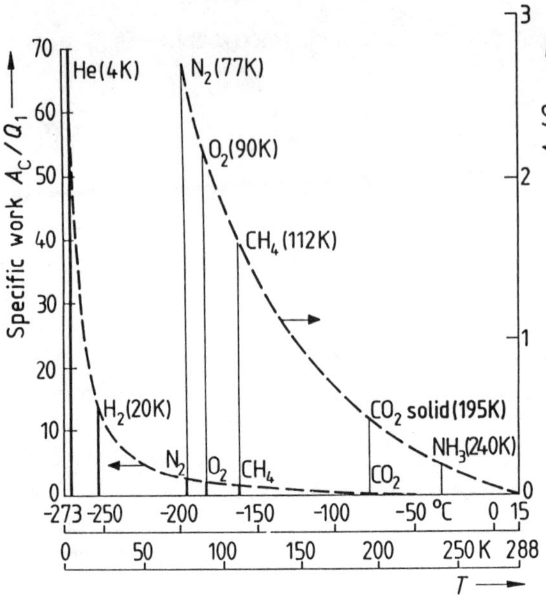

Fig. 6.2. Specific expenditure of work A_C/Q_1 in the Carnot cycle for refrigeration as a function of the refrigeration temperature T for an ambient temperature $T_{amb} = 288$ K, from [6.3]

falls again to the value T_1. The cycle is complete and now begins again. For the Carnot cycle we have

$$A_C = Q_1 \frac{T_2 - T_1}{T_1} \quad . \tag{6.7}$$

Figure 6.2 shows the specific expenditure of work in the Carnot cycle for the refrigeration as a function of the refrigeration temperature. The expenditure of work according to Fig. 6.2 represents a theoretical lower limit. In all actual refrigeration machines the expenditure of work is many times greater.

The refrigeration methods described in what follows all rely on gas expansion. According to the type of cyclic process applied the gas is either expanded isenthalpically through a throttle valve, or isentropically in an expansion machine. In continuous operation the work applied is used to compress the gas again. Besides the expansion processes a heat exchange is necessary between the cold gas at low pressure and the subsequent flow of warmer gas at high pressure. For this purpose counter-current heat exchangers or regenerators are used.

6.1.1 Joule-Thomson Expansion

If a real gas is expanded at constant enthalpy through a valve (throttle), the gas may – depending upon its initital state – be warmed or cooled. In the T, p diagram the region for warming ($\Delta T > 0$) and that for cooling ($\Delta T < 0$) are separated from each other by the inversion curve, see Fig. 6.3. With an operating point inside the inversion curve, i.e. with $\Delta T < 0$, it is called a Joule-Thomson expansion.

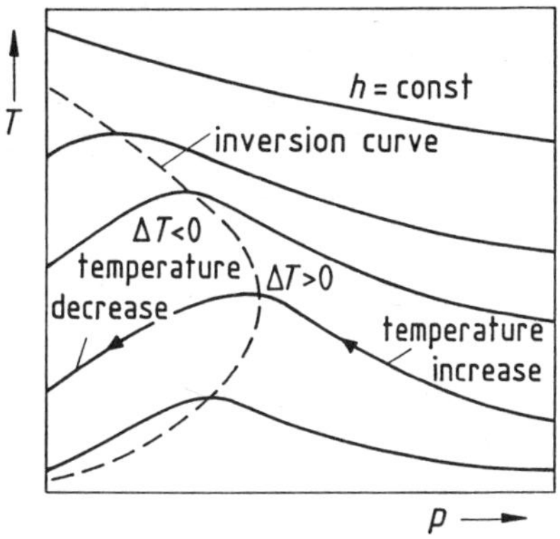

Fig. 6.3. Isenthalpic expansion of a real gas, from [6.2]

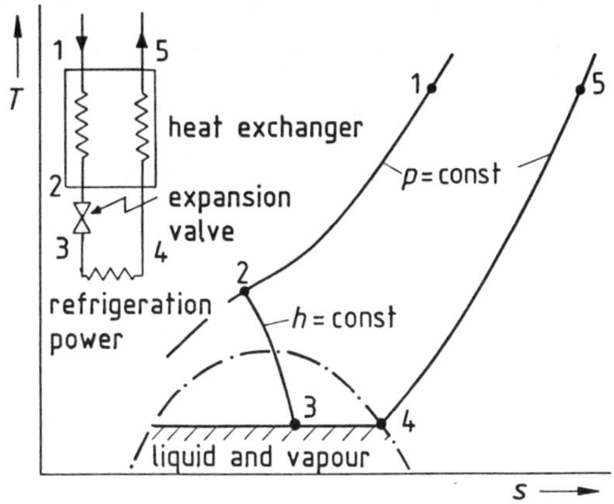

Fig. 6.4. Joule-Thomson expansion in the temperature-entropy (T, s) diagram

In the liquefaction of nitrogen low temperatures can be achieved by the Joule-Thomson expansion alone, since the inversion curve includes 293 K, but hydrogen and helium have to be precooled by other measures, because for them the ambient temperature lies outside the inversion curve. For the precooling one may consider the use of liquid nitrogen or isentropic expansion with the performance of work. The Joule-Thomson expansion then follows in the last cooling stage.

According to Fig. 6.4 constant temperatures can only be maintained if a part of the expanded gas is liquefied. In helium liquefaction this is in general 20 % of the gas. The non-liquefied portion of the cryogenic gas is conducted at low pressure to the lowest counterflow heat exchanger. In order to achieve a good efficiency, the temperature difference between the high pressure and the low pressure gas at the warm end of the lowest heat exchanger should be as small as possible. This is achieved by precooling to relatively low temperatures and the highest possible efficiency in the heat exchanger.

6.1.2 Expansion Machines

If gas in an expansion machine is expanded, doing work, this process is in the ideal case isentropic. Since it is also reversible it is always more efficient than a simple Joule-Thomson expansion. The isentropic expansion can moreover be applied above the inversion temperature. Figure 6.5 shows two isentropic expansion processes in the T, s diagram.

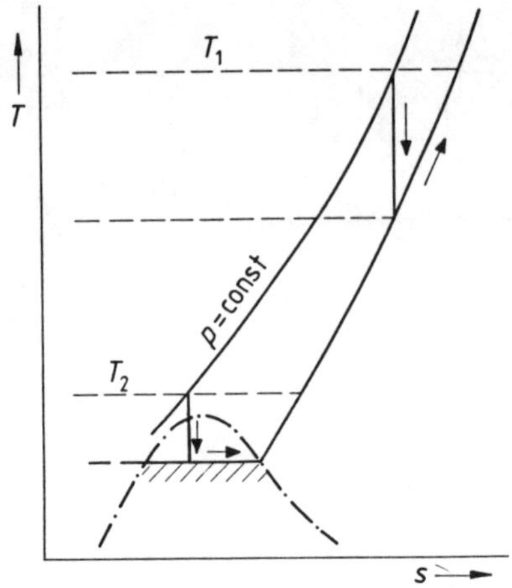

Fig. 6.5. Isentropic expansion in an expansion machine

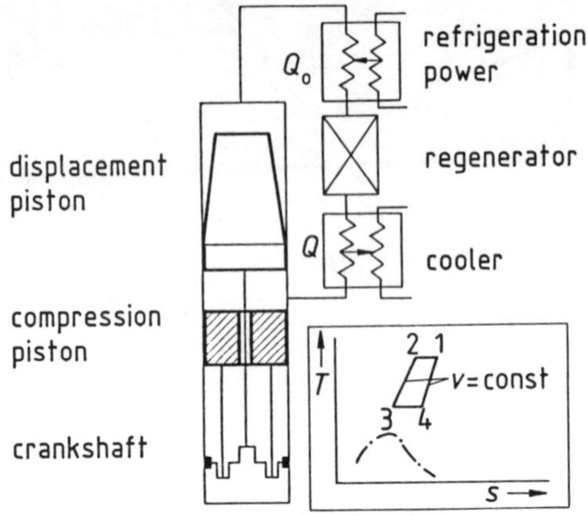

Fig. 6.6. Stirling method

6.1.3 Stirling Method

In this method for the production of low temperatures, shown in Fig. 6.6, a common cylinder contains a compression piston and a displacement piston. They are operated by a common crankshaft and driven in opposite directions. The working gas, usually helium, is contained in a closed system. The system includes a regenerator, which consists of a heat store with great thermal capacity and great heat transfer surface. Starting from state (1) the gas is compressed between the two pistons and delivers its compression heat in running through the cooler (2). Then it is cooled in the regenerator to about the temperature which is the target (3). Then there is an almost isothermal expansion between the displacement piston and the closed cylinder end (4). The expanded cold gas thereby gives up its store of cold to the outside. It then flows through the regenerator, which stores the residual usable coldness, and re-enters, warmed, into the compression chamber between the two pistons (1). The working gas is thus pushed to and fro between the compression chamber and the expansion chamber. It has even in the expanded state a relatively high pressure.

With two-stage Stirling refrigerators temperatures down to about 10 K can be achieved. This limit is imposed by the rapidly falling specific heat of the regenerator material at low temperatures. By connecting a Joule-Thomson cooling stage, however, one obtains a powerful and compact helium liquefier.

6.1.4 Gifford-McMahon Method

In the Gifford-McMahon method, like the Stirling method, the working gas is pushed to and fro through a regenerator by a displacement piston. As shown in Fig. 6.7, however, the compression is carried out in a compressor at room temperature separated from the cooling system. In this process no work is done by the system, but a quantity of heat corresponding to the quantity of coolness produced is discarded to the exterior. In the same way as coolness is produced by gas expansion on the cold side of the displacement piston, heat is produced by compression of the gas on the room temperature side of the displacement piston. The gas emerging from the system therefore has a higher temperature than the gas entering the system.

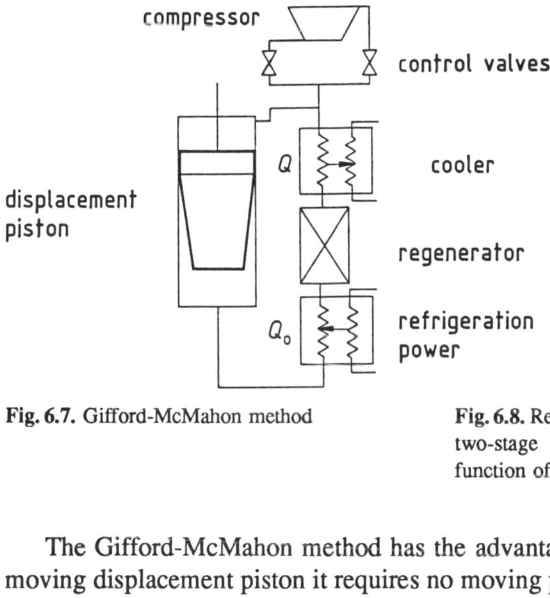

Fig. 6.7. Gifford-McMahon method

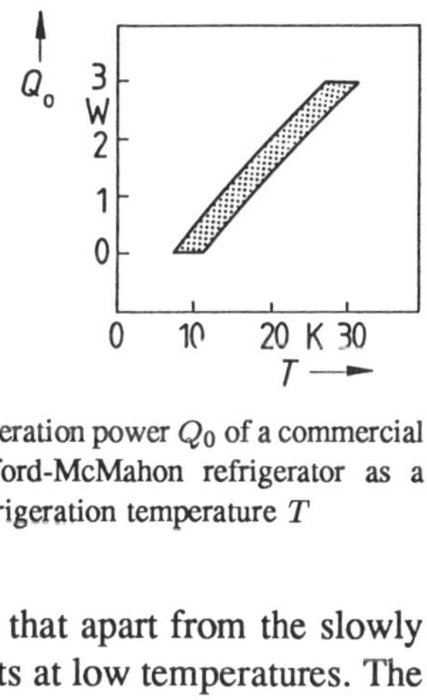

Fig. 6.8. Refrigeration power Q_0 of a commercial two-stage Gifford-McMahon refrigerator as a function of refrigeration temperature T

The Gifford-McMahon method has the advantage that apart from the slowly moving displacement piston it requires no moving parts at low temperatures. The compressor can be operated separately from the actual cool system. The control valves also are at room temperature.

As shown in Fig. 6.8, a two-stage Gifford-McMahon refrigerator enables one to achieve temperatures down to about 10 K.

6.2 Cooling in Liquid Bath Cryostats

If specimens or electronic components are to be cooled by two-stage gas coolers to temperatures above 10 K, one establishes good thermal conduction between them and the low temperature stage of the refrigerator. In addition one places a radiation screen round the other stage. With such a system installed in a vacuum chamber, temperatures between 10 K and 300 K can be applied by using a heater at the specimen mounting. The vacuum chamber can contain windows for the input or output of radiation. The important advantage of this cryostat lies in that it makes possible experiments at low temperatures without a cryogenic fluid. In a refrigerator with a Joule-Thomson stage the liquid-gas mixture issuing from it is led in its entirety to the user and the cold gas from this is fed again into the lowest counterflow heat exchanger of the cooling installation.

On the other hand it is also possible to pour the liquid refrigerant, either produced on the spot or brought there in portable containers, into the cryostat in which the object to be cooled is also located. In the case of helium the gradually evaporating gas is usually collected again and led into a condenser.

Cryostats in the form in which they were invented by Dewar around the turn of the century, and later named after him, are still obtainable even today. The principle consists of a vaccum-insulated glass flask. For liquid nitrogen (LN_2) one uses it in its simple form, for liquid helium (LHe) one uses two coaxial Dewar flasks, the outer one cooled with LN_2 and serving as a radiation screen for the inner flask filled with LHe. Depending upon the application, the walls of the flask are totally silvered or furnished with a viewing strip. Glass as manufacturing material has the advantages of low thermal conductivity, low gas emission and easy fabrication properties, but also the disadvantages of fragility, limitation in dimensions and permeability for helium. The latter has the result that the vacuum mantle of a LHe flask has to be re-evacuated after a certain period of use.

These disadvantages are avoided in cryostats made of metal. Construction from metal also allows small distances to be maintained between the outer mantle, radiation screen and inner wall, so that there is a favourable ratio of usable inner space to total volume. One differentiates between cryostats which are of completely welded construction and those which also incorporate flange fastenings and can therefore be disassembled and are universal in application. Transport cryostats are usually of metal. In the case of helium and with small objects to be refrigerated it is often appropriate to refrigerate these directly in a small transport container (capacity 20 to 50 l).

In the liquid bath cryostat the temperature of the experiment is determined by the temperature of the refrigerant. Within limits it can be varied by changes of pressure. By low pressure (pumping down) one can obtain in this way temperature ranges of 1 to 4.2 K for LHe, 13.9 to 20.4 K for LH_2 and 63 to 77 K for LN_2.

To lower the pressure of helium one must take account of the following. At 2.18 K and 5×10^3 Pa, the lambda point, normal liquid He I passes into the superfluid phase He II. This phase transformation is associated with abrupt changes of several physical properties. Thus He II has a thermal conductivity which is about six powers of ten greater than that of He I. It thus exceeds the thermal conductivity of the purest metals by almost two orders of magnitude.

He II has also an extremely small viscosity. Under the smallest pressure differences it flows through cracks and pores down to about 10^{-5} cm. Moreover it creeps up the walls of the flask in the direction of higher temperatures.

According to its origin helium consists of a variable mixture of two stable isotopes with masses 4 (^4He, abundance 99.99987 %) and 3 (^3He, abundance 0.00013 %). With a pressure of 10^5 Pa the boiling points of ^4He and of ^3He are 4.2 K and 3.2 K, respectively. With ^3He lower temperatures can be obtained than with ^4He.

It should be noted in this connection that the constituents of the air at temperatures < 30 K are solid. They therefore can become disturbing impurities which cause stoppages of pipes, valves, etc.

6.3 Temperature Measurement Techniques

The thermodynamic temperature scale is standardised by taking the triple point of water at $T = 273.16$ K as the fixed point. Since this triple point lies at 0.01 °C, the conversion from the Kelvin scale to the Celsius scale is

$$T(\text{in K}) = t(\text{in}\,^\circ\text{C}) + 273.15 \quad . \tag{6.8}$$

For the realisation of the thermodynamic scale of temperature one uses primary thermometers, such as the helium gas thermometer. In technical applications one uses secondary thermometers, which are obtainable from the trade together with the appropriate measuring instruments. Table 6.1 from [6.1] gives a review of secondary thermometers for low temperatures.

In resistance thermometers the electrical resistance is measured as an indicator of the temperature. More specifically it is measured:

— by the voltage drop U with a known constant current I using a digital voltmeter or a compensation recorder,
— potentiometrically with compensator and standard resistance or
— with a resistance measuring bridge.

The resistance of the connecting wires is here eliminated by a four-lead arrangement. In spite of the greater expense the measurement using alternating current has the advantage that thermovoltages have no effect. Table 6.1 lists various commonly used resistance materials. Here thermistors are semiconductor elements made from metal oxides by sintering, which have a falling $R(T)$ characteristic.

Temperature probes in the form of semiconductor diodes are based on the temperature dependence of the forward voltage U_d at a constant current I. The measurement of U_d is made using a digital voltmeter with an impressed current I.

Table 6.1. Secondary Thermometers for Low Temperatures, from [6.1]

	Range in K	Sensor volume in cm^3	Meas. power in μW	Meas. error in mK	Reproducibility in mK
Resistance thermometers:					
Platinum	20...600	< 1	1...100	10*	1
Rhodium + 0.5 at. % iron	4...300	< 1	1...100	10*	1
Germanium	10^{-2}...100	10^{-1}	< 1	10*	0.1
Carbon, encapsulated	10^{-2}...50	10^{-1}	1...10	10*	1
Carbon glass	10^{-2}...300	10^{-1}	1	10*	1
Thermistor	1...400	10^{-2}	< 10	10*	1
Semiconductor diodes:					
Gallium arsenide	1...400	10^{-2}	< 10	10*	1
Silicon	1...400	10^{-2}	< 10	10*	1
Thermo-couples:					$\delta T/T$ in %
Copper/constantan	50...800	10^{-4}	$\simeq 0$	**	0.5
Copper/gold + 2.1 at. % cobalt	20...300	10^{-4}	$\simeq 0$	**	1
Chromel/gold + 0.03 to 0.07 at. % iron	1...500	10^{-4}	$\simeq 0$	**	0.5
Silver°/gold + 0.03 at. % iron	1...20	10^{-4}	$\simeq 0$	**	0.5

° silver with 0.37 at.% gold
* dependent on the calibration method
** No absolute measurement

Thermo-couples consist of two different materials, at whose contact there is a temperature-dependent thermo-electric voltage. The sum of the thermo-electric voltages is measured with a compensation recorder or a digital voltmeter. It is necessary, however, to have a constant and well-known reference temperature T_0, which one obtains with Peltier elements from triple point cells, cold baths or freezing-point thermostats.

These and other methods of measurement are described in detail in [6.1].

In the construction of temperature probes it is necessary to fulfil the following conditions: good thermal contact with the cold surface, screening against thermal radiation and meticulous avoidance of incoming or outgoing heat currents via the measuring wires.

The calibration of the measuring probes is usually carried out by means of reference temperature probes. In the temperature range below 50 K one uses a calibrated germanium temperature probe, and in the range above 50 K a calibrated platinum temperature probe. From the experimental calibration data one can then use the method of least squares by a computer program to calculate the coefficients of a polynomial representation of $R(T)$; from this one prepares the $R(T)$ table for later use.

6.4 Materials

The variaton of the physical properties of different materials with decreasing temperature is very diverse.

The thermal conductivity of pure metals and other pure materials at first increases below 100 K. At about 20 to 10 K it passes through a maximum and at lower temperatures the thermal conductivity decreases again. Alloys, plastics and glasses, on the other hand, show a continual decrease in thermal conductivity, with the plastics and glasses lying one to two orders of magnitude below the alloys and having about the same thermal conductivity as cold liquids. Of the electrical insulators, sapphire and diamond have very high thermal conductivities. They lie at more than four orders of magnitude above those of glass and plastics.

Also the thermal expansion of the materials is very varied. This must be taken into account in all design processes, in order to avoid damage by differential contraction. Figure 6.9 shows the behaviour of the linear expansion coefficients of different materials as a function of the temperature. For most materials the

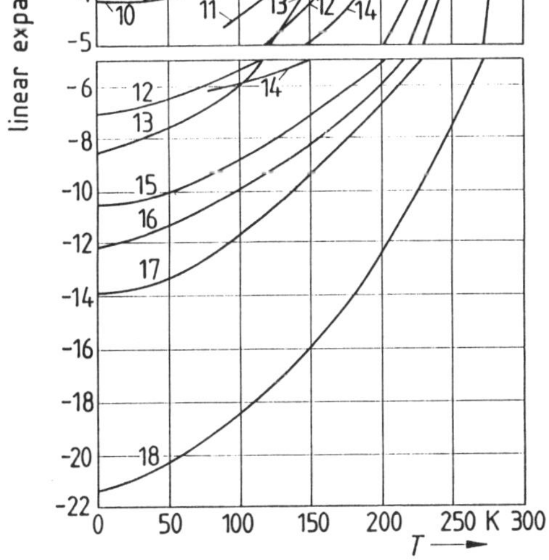

Fig. 6.9. Integral thermal linear expansion of a few materials, from [6.1]. (1) Invar, (2)Pyrex glass, (3) unalloyed steel, (4) nickel, (5) Contracid, (6) stainless steel, (7) copper, (8) nickel silver, (9) brass, (10) aluminium, (11) soft solder, (12) indium, (13) mercury, (14) ice, (15) Araldite, (16) Plexiglass, (17) Nylon, (18) Teflon

order of magnitude of α lies between $10^{-5}\,K^{-1}$ and $10^{-6}\,K^{-1}$. An important exception is Invar. This material is particularly appropriate for joining glass and metal. The thermal expansion coefficient of plastics can be matched to the thermal expansion of nearly all materials by suitable filler materials.

Whereas many materials become very brittle and therefore useless as the temperature falls, the strength of other materials increases. The elasticity decreases. This is of particular importance for sealing materials. Instead of the usual elastic sealing materials, therefore, at low temperatures one uses metals which are still ductile, especially indium, and other sealants of very special construction.

6.5 Cooling Systems for Microwave Receivers

In this section a few examples of cooling systems for microwave receivers will be described.

Figure 6.10 shows the design of a mixer with Josephson point contacts from [6.5]. The cooling takes place in a bath cryostat using liquid helium, where the mixer is immersed in the helium. The inner Dewar flask is cooled by liquid nitrogen on its outside. The precooling of the mixer is also achieved by liquid nitrogen, which is pumped out before filling with liquid helium. A helium filling lasts for about three or four hours. In order to keep the heat conduction low, the coaxial cable for the d.c. and intermediate frequency connections is made from material with low heat conduction. The waveguide, which feeds in the receiver signal and the local oscillator signal, has its wall thickness minimised by etching. This also serves to keep the heat conduction and hence the evaporation rate of the liquid helium low. Also the waveguide material itself should be a poor heat conductor. Tubes of coin silver or even of stainless steel may be used, and they are then thinly silvered on the inside to increase the electrical conductivity.

In the experimental set-up of Fig. 6.10 a waveguide sliding short and the contact mechanism for the Josephson point contact can be adjusted from outside. For mechanical operation nylon rods were used.

The hybrid cryostat from [6.7], shown in Fig. 6.11, is in regular use for SIS mixers. It uses an industrial two-stage Stirling refrigerator, in which radiation screens at 70 K and 15 K are introduced at the two cooling stages. There is a reservoir of liquid helium inside the 15 K screen. The consumption of liquid helium is therefore small and the lifetime of the cryostat is kept high. The liquid helium is pumped from the reservoir through the mixer chamber. A temperature below 4 K is thus maintained. The helium gas cools the mixer housing and the SIS junction contained therein. The temperature of the junction is measured with a thermometer not shown in Fig. 6.11. The heat generated in the junction by the d.c. and the local oscillator power is carried off by the helium gas. The excess of temperature of the SIS junction, relative to the mixer housing, is estimated to be only a fraction of a degree. The temperature of the SIS junction can therefore be determined with sufficient accuracy. The receiver signal and the local oscillator

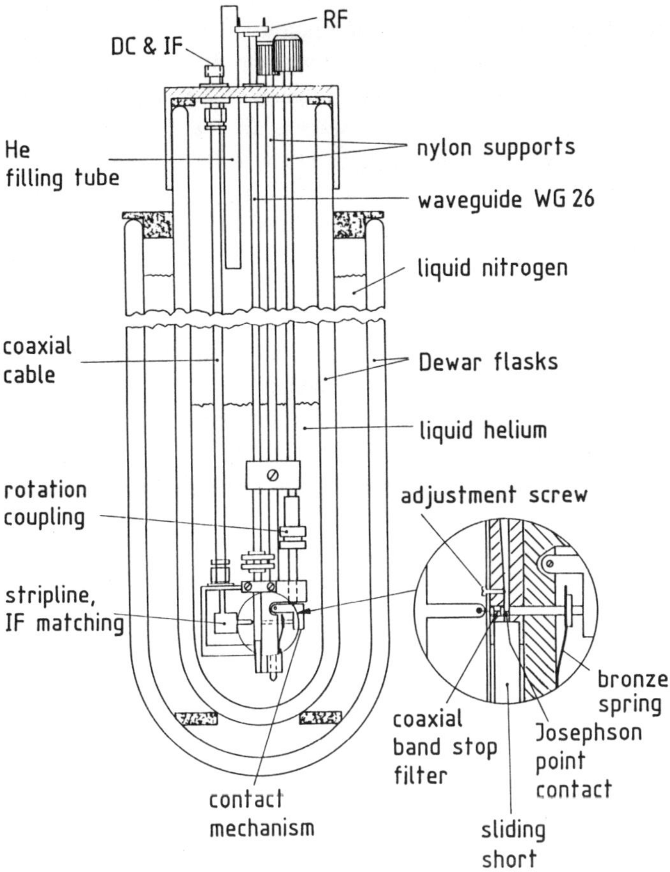

Fig. 6.10. Design for mixer with Josephson contact, from [6.6]

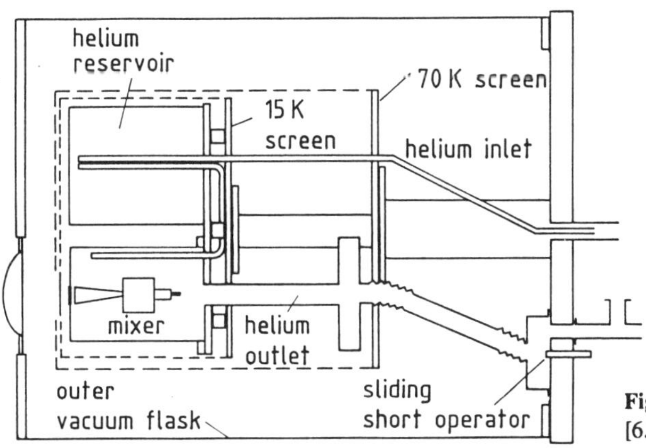

Fig. 6.11. Hybrid cryostat from [6.7] for SIS mixer

signal are led into the mixer through a system of windows. The window at room temperature is a lens of Teflon, fitted into the outer vacuum flask of the cryostat. At about 100 GHz the loss through this lens is estimated at 0.14 dB. Behind the lens there is a fluorogold window in the 15 K screen. This window acts as an infrared absorber and is transparent to frequencies below 1 THz. Finally there is a helium-proof quartz window in the mixer chamber. This window is matched on both sides for radiofrequencies with a layer of Teflon. The total attenuation of the lens and window system is about 0.15 dB.

The intermediate frequency output of the mixer is connected over a coaxial cable of stainless steel with the first stage of the intermediate frequency amplifier. This is an FET amplifier in an intermediate chamber at about 15 to 20 K. This amplifier chamber is ventilated with helium gas which comes from the mixer chamber. The helium reservoir and the mixer chamber are held by supports which have a low thermal conductivity. In this way the mixer can be operated for more than 24 hours at temperatures below 4 K on one filling of a litre of liquid helium.

In both of the systems described above the liquid helium is not in a closed cycle. The operating time is therefore limited. A refill of liquid helium is needed. For operation down to below 4.5 K refrigerators with closed cycles are obtainable from the trade. For the operation of SIS mixers, however, temperature reduction to 2.5 K can bring significant advantages. A special refrigerator for the receiver of a radio-telescope has therefore been developed, see [6.8].

Again it consists first of a two-stage Stirling refrigerator, with which the helium gas is precooled to 18 K with a pressure of 1.2 MPa. Then it is expanded in a Joule-Thomson stage to about 10 kPa. A vacuum pump and a compressor bring the gas back to its initial pressure. The whole refrigerator fits in the receiver chamber of a small radio-telescope and works largely independently of the orientation. After a cooling time of 12 to 24 hours a cooling power of 200 to 250 mW is available at 2.8 K.

List of Principal Symbols

A	area; magnetic vector potential
B	band width; magnetic flux density
C	capacitance
C'	capacitance per unit length
d	electrode separation distance
D	density of states
e	elementary charge
E	electric field strength
f	frequency
f_{c}	cut off frequency
$f_{\mathrm{p}},\ f_{\mathrm{p}}'$	plasma frequency
F	noise figure
$f_{\mathrm{erl,lr}}$	electron currents from right to left, left to right
G	gain; conductance
G_I	current gain
G_{m}	maximal gain
h	height; Planck's constant; enthalpy
$\hbar$	Planck's constant divided by 2π
H	magnetic field strength
$\mathcal{H}(x)$	Hilbert transform
$i,\ I$	electrical current
j	imaginary unit
J	current density
J_A	surface current density
$J_n(x)$	Bessel function of the nth order
k	wave number; integer
k_{B}	Boltzmann constant
K_{f}	flicker noise constant
$K_0(x)$	modified Bessel function of the order zero
l	length, free path length; integer
L	inductance
L'	inductance per unit length
m	electron mass; integer
M	magnetisation
n	integer; charge carrier density
n_m	demagnetisation factor

N	integer
NEP	noise equivalent power
p	momentum; pressure
P	power
P'	power density
P_i	noise power normalised to the resistance
q, Q	electrical charge; quantity of heat
r	magnitude of position vector
R	electrical resistance
R'	resistance per unit length
R_d	differential resistance; dynamic resistance
R_i	current sensitivity
R_s	surface resistance
R_{sg}	mean resistance below the energy gap voltage
R_u	voltage sensitivity
s	path length; entropy
S_i	noise power density normalised to the resistance
SCF	strong coupling factor
t	time
T	absolute temperature
T_c	critical temperature
$T_{e,M,R}$	equivalent noise temperature of an amplifier, mixer, receiver
u, U	voltage; internal energy
u_i	induced voltage
U_gap	energy gap voltage
v	velocity; volume
v_m	mean velocity
w	width of the tunnel junction
W	energy
x	Cartesian coordinate
y	Cartesian coordinate
Y	admittance
z	Cartesian coordinate
Z	wave impedance
Z_s	surface impedance

Greek Symbols

α	attenuation constant
β	phase constant
β_c	McCumber parameter
Γ	noise parameter
δ	line width in the frequency spectrum
δ_a	anomalous skin depth
δ_c	classical skin depth
Δ	difference; 2Δ energy gap in the superconductor

ε	permittivity
ε_r	relative permittivity
ε_0	permittivity of vacuum
η	mixer parameter
θ	phase angle
κ	Ginzburg-Landau parameter
λ	penetration depth of the magnetic field into the superconductor
λ_L	London penetration depth
μ	permeability
μ_r	relative permeability
μ_0	permeability of vacuum
ξ_{co}	coherence length of Cooper pairs
ξ_{GL}	Ginzburg-Landau coherence length
ϱ	resistivity
σ	conductivity
φ	phase difference
Φ	magnetic flux
Φ_0	flux quantum
χ_m	magnetic susceptibility
Ψ	wave function
ω	angular frequency
ω_c	characteristic angular frequency of a Josephson junction
Ω	normalised frequency

Indices

A	area, surface
av	available
B	image frequency
c	critical
co	Cooper pair
dc	d.c. current
DSB	double side band
e	electron; external
F	Fermi
g, gap	energy gap
G	generator
h	hole
H	auxiliary frequency
I	intermediate frequency
in	input
J	Josephson
K	periodic circuit; short-circuit
l	left
L	load; conduction band
LO	local oscillator

max	maximal
min	minimal
M	mixer
n	normally conducting
N	noise
out	output
r	right
RF	radio frequency
s	superconducting
sat	saturated
S	signal
SSB	single side band
v	lossy
V	amplifier
W	in phase contribution
0	at the work point

Special Symbols

$(\hat{\ })$	amplitude
$(\underline{\ })$	phasor
$(\bar{\ })$	mean value
$(^{*})$	conjugate complex
Δ	difference
$(\dot{\ })$	derivative with respect to time
Bold Type (or $(\vec{\ })$ in the figures)	vector, matrix

Literature

Chapter 1

1.1 Rose-Innes, A.C.; Rhoderick, E.H.: *Introduction to Superconductivity*, 2nd edn. Oxford etc.: Pergamon Press 1979

1.2 Buckel, W.: *Supraleitung (Superconductivity)*, 2nd ed. Weinheim: Physik Verlag 1977

1.3 Van Duzer, T, Turner C.W.: *Principles of Superconductive Devices and Circuits*, New York: Elsevier North Holland 1981

1.4 Gallop, J.C.: *Use of persistent supercurrents in SQUID current stabilizers and their application to a resistivity measurement on niobium*, J. Phys. D. 9 (1976) 2111–2115

1.5 Meissner, W.; Ochsenfeld, R.: *Ein neuer Effekt bei Eintritt der Supraleitung (A new effect at the onset of superconductivity)*, Naturwissenschaften 21 (1933) 787

1.6 Lautz, G.: *Elektromagnetische Felder (Electromagnetic Fields)*, 3rd ed. Stuttgart: Teubner 1985

1.7 London, F.; London, H.: *The Electromagnetic Equations of the Supraconductor*, Proc. Roy. Soc. London (A) (1935) 71–88

1.8 Bardeen, J.; Cooper, L.N.; Schrieffer, J.R.: *Theory of Superconductivity*, Phys. Rev. 108 (1957) 1175

1.9 Unger, H.-G.; Schultz, W.; Weinhausen, G.: *Elektronische Bauelemente und Netzwerke I (Electronic Components and Networks I)*, 3rd ed. Braunschweig (Brunswick): Vieweg & Sohn 1979, reprint 1985

1.10 Unger, H.-G.: *Elektromagnetische Theorie für die Hochfrequenztechnik (Electromagnetic Theory of Radio Frequency Technique)*, Part I, 2nd ed. Heidelberg: Hüthig 1988

1.11 Hinken, J.H.; Pöpel, R.: *Gekühlte Mikrostreifenleitungen für Josephson-Spannungsnormale (Low Temperature Microstrip Conductors for Josephson Voltage Standards)*, Nachrichtentechn. Zeitschr. Archiv 5 (1983) 199–220

1.12 Mattis, D.C.; Bardeen, J.: *Theory of the anomalous skin effect in normal and superconducting metals*, Phys. Rev. 111 (1958) 412

1.13 Kautz, R.L.: *Miniaturisation of Normal-State and Superconducting Striplines*, J. of Res. Nat. Bureau of Standards 84 (1979) 247–259

1.14 Pöpel, R.: *Auswertung der Mattis-Bardeen-Theorie und Messungen an supraleitenden Mikrostreifenleitungen (Evaluation of the Mattis-Bardeen Theory and Measurements on Superconducting Microstrip Transmission Lines)*, Dissertation, Technische Universität Braunschweig (Brunswick) 1986

1.15 Schiff, L.I.: *Quantum mechanics*, 3rd ed.; Tokyo: McGraw-Hill 1968

1.16 Silsbee, F.B.: *A note on electrical conduction in metals at low temperatures*, J. Wash. Acad. Sci. 6 (1976) 597

1.17 Ginzburg, V.L.; Landau, L.D.: ..., Zh. Eksp. Teor. Fiz. 20 (1950) 1044 (in Russian)

1.18 Pippard, A.B.: *Metallic Conduction at High Frequencies and Low Temperatures* in *Advances in Electronics*, L. Norton, Academic Press, New York (1954)

1.19 Huebener, R.P.: *Magnetic Flux Structures in Superconductors*, Berlin, Heidelberg, New York: Springer 1979

Chapter 2

2.1 Giaever, I.: *Energy gap in superconductors measured by electron tunneling*, Phys. Rev. Lett. 5 (1960) 147–148

2.2 Solymar, L.: *Superconductive Tunneling and Applications*, London: Chapman and Hall, 1972. See also 1.3

2.3 Tucker, J.R.; Feldman, M.J.: *Quantum detection at millimeter wavelength*, Rev. Mod. Phys. 57 (1985) 1055–1113

2.4 Barone, A.; Paterno, G.: *Physics and Applications of the Josephson Effect*, New York etc.: John Wiley & Sons 1982

2.5 Hartfuss, H.J.; Gundlach, K.H.: *Video Detection of mm-Waves via Photon-Assisted Tunneling between two Superconductors*, Int. J. Infrared Millimeter Waves 2 (1981) 809

2.6 Unger, H.-G.; Schultz, W.; Weinhausen, G.: *Elektronische Bauelemente und Netzwerke II (Electronic Components and Networks II)*, 3rd ed. Braunschweig (Brunswick): Vieweg & Sohn 1981

2.7 Held, D.N.; Kerr, A.R.: *Conversion Loss and Noise of Microwave and Millimeter-Wave Mixers: Part 1 – Theory*, IEEE Trans. Microwave Theory Techn. MTT-26 (1978) 49–55

2.8 Torrey, H.C.; Whitmer, A.C.: *Crystal Rectifiers*, Sec. 5.3, New York: McGraw-Hill Book Company 1949

2.9 Woody, D.P.; Miller, R.E.; Wengler, M.J.: *85 – 115-GHz Receivers for Radio Astronomy*, IEEE Trans. Microwave Theory Techn. MTT-33 (1985) 90–95

2.10 Hartfuss, H.J.; Tutter, M.: *Numerical Design Calculations of a mm-wave Mixer with SIS Tunnel Junction*, Int. J. Infrared Millimeter Waves 4 (1983) 993–1014

2.11 Hartfuss, H.J.; Tutter, M.: *Minimum Noise Temperature of a Practical SIS Quantum Mixer*, Int. J. Infrared Millimeter Waves 5 (1984) 717–734

2.12 Blundell, R.; Ibruegger, J.; Gundlach, K.H.; Blum, E.J.: *SIS-mixer receiver with single and array tunnel junctions for the 140 GHz to 170 GHz range*, Proc. 14th Europ. Microwave Conf., Liège (1984) 581–586

2.13 McGrath, W.R.; Richards, P.L.; Smith, A.D.; van Kempen, H.; Batchelor, R.A.; Prober, D.E.; Santhanam, P.: *Large gain, negative resistance, and oscillations in superconducting quasiparticle heterodyne mixers*, Appl. Phys. Lett. 39 (1981) 655–658

2.14 Smith, A.D.; McGrath, W.R.; Richards, P.L.; van Kempen, H.; Prober, D.; Santhanam, P.: *Negative resistance and conversion gain in SIS-mixers*, Physica 108B (1981) 1367–1368

2.15 Räisänen, A.V.; Crété, D.G.; Richards, P.L.; Lloyd, F.L.: *Wide-band ultra low noise mm-wave mixers with a single tuning element*, Proc. 16th Europ. Microwave Conf., Dublin (1986) 252–257

2.16 Ibruegger, J.; Cartel, M.; Blundell, R.: *A Superconducting Mixer Receiver for the Frequency Range from 125 GHz to 175 GHz*, MIOP '87 Kongressunterlagen (Congress Proceedings), Vol. 1, Hagenburg: NETWORK GmbH, 1987, Contribution 1B–4

2.17 Crété, D.-G.; McGrath, W.R.; Richards, P.L.; Lloyd, F.L.: *Performance of Arrays of SIS Junctions in Heterodyne Mixers*, IEEE Trans. Microwave Theory Techn. MTT-35 (1987) 435–440

2.18 D'Addario, L.R.: *An SIS-mixer for 90–120 GHz with gain and wide bandwidth*, Int. J. Infrared Millimeter Waves 5 (1984) 1419–1442

2.19 D'Addario, L.R.: Noise Parameters of SIS-Mixers, IEEE Trans. Microwave Theory Techn. MTT-36 (1988) 1196-1206

Chapter 3

3.1 Josephson, B.D.: *Possible new effects in superconductive tunneling*, Phys. Lett. 1 (1962) 251–253

3.2 Van Duzer, T.; Turner, C.W.: *Principles of superconductive devices and circuits*, New York: Elsevier North-Holland 1981

3.3 Barone, A.; Paterno, G.: *Physics and applications of the Josephson effect*, New York etc.: John Wiley & Sons 1982

3.4 Solymar, L.: *Superconductive tunneling and applications*, London: Chapman and Hall 1972

3.5 Feynman, R.P.; Leighton, R.B.; Sands, M.: *The Feynman lectures on physics*, Vol. III. Reading Massachusetts: Adison-Wesley 1965

3.6 Ambegaokar, V.; Baratoff, A.: *Tunneling between superconductors*, Phys. Rev. Lett. 10 (1963) 486–489. Errata: Phys. Rev. Lett. 11 (1963) 104

3.7 Kanter, H.; Vernon, F.L.: *High-frequency-response of Josephson point contacts*, J. Appl. Phys. 43 (1972) 3174

3.8 McCumber, D.E.: *Effect of ac impedance on dc voltage-current characteristics of superconductor weak-link junctions*, J. Appl. Phys. 39 (1969) 3113–3118

3.9 Scott, W.C.: *Hysteresis in the dc switching characteristics of Josephson junctions*, Appl. Phys. Lett. 17 (1970) 166–169

3.10 Klinger, M.: *Berechnung von Nullstromstufen an Josephon-Tunnelelementen unter Berücksichtigung der Gleichrichterwirkung am Quasiteilchenwiderstand (Calculation of Zero-current Steps in Josephson Tunnel Junctions taking account of the Rectification Effect on the Quasiparticle Resistance)*, Entwurf am Institut für Hochfrequenztechnik (Note of the Institute of High Frequency Engineering) Technische Universität Braunschweig (Brunswick) 1984

3.11 Falco, C.M.; Parker, W.H.; Trullinger, S.E.; Hansma, P.K.: *Effect of thermal fluctuations on the I-V characteristics of a highly damped Josephson junction*, Phys. Rev. B. 10 (1974) 1867

3.12 Zinke-Brunswig: *Lehrbuch der Hochfrequenztechnik (Textbook of Radio Frequency Technique)*, Vol. 2, Berlin etc.: Springer 1987

3.13 Grimes, C.C.; Shapiro, S.: *Millimeter-wave mixing with Josephson junctions*, Phys. Rev. 169 (1968) 397–406

3.14 Harris, E.P.; Laibowitz, R.B.: *Properties of superconducting weak links prepared by ion implantation and by electron beam lithography*, IEEE Trans. Magn. 13 (1977) 724–730

3.15 Russer, P.: *Influence of microwave radiation on current-voltage characteristic of superconducting weak links*, J. Appl. Phys. 43 (1972) 2008–2010

3.16 Kautz, R.L.; Monaco, R.: *Survey of chaos in the rf-biased Josephson junction*, J. Appl. Phys. 57 (1985) 875–889

3.17 Noeldeke, C.; Gross, R.; Bauer, M.; Reiner, G.; Seifert, H.: *Experimental survey of chaos in the Josephson effect*, J. Low Temp. Phys. 64 (1986) 235–268

3.18 Hinken, J.H.; Brunk, G.; Cui, G.-Y.; Niemeyer, J.: *Measured flatness of microwave-induced steps at Josephson tunnel junctions with near-zero-current bias*, IEEE Trans. Instr. Meas. IM-31 (1982) 223–226

3.19 Unger, H.-G.: *Elektromagnetische Wellen auf Leitungen (Electromagnetic waves on transmission lines)*, Heidelberg: Hüthig 1980

3.20 Rebbi, C.: *Solitons*, Scientific American, 240, No. 2 (Feb. 1979), 76–91

3.21 Klein, U.; Hinken, J.H.: *Current density distribution in one-dimensional Josephson tunneling junctions at microwave induced constant voltage steps*, Ext. Abstr. 1987 Intern. Supercond. Electronics Conf. (ISEC '87), Tokyo, 77–80

3.22 Clarke, J.: *Advances in SQUID magnetometers*, IEEE Trans. Electron Devices, ED-27 (1980) 1896–1908

3.23 Hahlbohm, H.D.; Lübbig, H. (Eds.): *SQUID '85 Superconducting Quantum Interference Devices and their Applications*, Berlin: de Gruyter 1985

3.24 Romani, G.L.; Williamson, S.J.; Kaufmann, L.: *Biomagnetic instrumentation*, Rev. Sci. Instr. 53 (1982) 1815–1845

3.25 Beha, J.: *Digitale Speicherzellen mit Josephson-Kontakten (Digital Storage Cells with Josephson Junctions)*, Berlin: VDE-Verlag 1981

3.26 Ohta, H.: *A self-consistent model of the Josephson junction, Superconducting Quantum Interference Devices and their Applications: SQUID*, H.D. Hahlbohm & H. Lübbig (Eds.); Berlin, New York: de Gruyter 1977, 35–49

Chapter 4

4.1 Kose, V.: *Recent Advances in Josephson Voltage Standards*, IEEE Trans. Instr. Meas. IM-25 (1976) 483–489

4.2 Quinn, T.J.: News from the BIPM, Metrologia 26 (1989) 69-74

4.3 Niemeyer, J.; Sakamoto, Y.; Vollmer, E.; Hinken, J.H.; Shoji, A.; Nakagawa, H.; Takada, S.; Kosaka, S.: *Nb/Al-oxide/Nb and NbN/MgO/NbN tunnel junctions in large series arrays for voltage standards*, Jpn. J. Appl. Phys. 25 (1986) L343–L345

4.4 Vollmer, E.: *Berechnung planarer Mikrowellenschaltungen für Spannungsnormale mit in Reihe geschalteten Josephson-Elementen (Calculation of planar microwave circuits for voltage standards using Josephson junctions connected in series)*, Dipl.-Arbeit am Institut für Hochfrequenztechnik (Diploma-Thesis of the Institute of High Frequency Engineering) Technische Unversität Braunschweig (Brunswick) 1984

4.5 Kautz, R.; Costabile, G.: *A Josephson voltage standard using a series array of 100 junctions*, IEEE Trans. Mag. MAG.-17 (1981) 780–783

4.6 Greiner, J.H. et al.: *Fabrication process for Josephson integrated circuits*,IBM J. Res. Develop. 53 (1982) 326–336

4.7 Baker, J.M.; Magerlein, J.H.: *Tunnel barriers on Pb-In-Au alloy films*, J. Appl. Phys. 54 (1983) 2556–2568

4.8 Hinken, J.H.; Niemeyer, J.: *Supraleitende integrierte Millimeterwellenschaltung mit 1474 Josephsonelementen für präzise Gleichspannungen bis zu 1,2 Volt (Superconducting integrated millimetrewave circuit with 1474 Josephson junctions for precise d.c. voltages up to 1.2 volts)*, Kleinheubacher Berichte Nr. 28, 1985, ISBN 0343-5725, 81–95

4.9 Unger, H.-G.:*Elektromagnetische Wellen auf Leitungen (Electromagnetic waves on transmission lines)*, 2nd ed. Heidelberg: Hüthig 1986, Section 4.3

4.10 Kircher, C.J.; Lahiri, S.K.: *Properties of $AuIn_2$ resistors for Josephson integrated circuits*, IBM J. Res. Develop. 24 (1980) 235–242

4.11 Vollmer, E.; Hinken, J.H.; Niemeyer, J.; Meier, W.: *Stable standard voltages of 1 volt by improved design of superconducting MMICs with 1440 Josephson junctions*, 15th Europ. Microw. Conf. (1985) 881

4.12 Niemeyer, J.; Grimm, L.; Meier, W.: *Stable Josephson reference voltages between 0.1 and 1.3 V for high-precision voltage standards*, Appl. Phys. Lett. 47 (1985) 1222–1223

4.13 Hinken, J.H.; Niemeyer, J.; Pöpel, R.: *E-band transformer from waveguide to superconducting, low-impedance, antipodal fin-line*, Nachrichtentechn. Zeitschr. Archiv 8 (1986) 215–222

4.14 Ljapin, V.L.; Hinken, J.H.: *Rigorous analysis of the antipodal fin-line with extremely thin isolating layer*, Proc. URSI Intern. Symp. Budapest (1986) 810–812

4.15 Niemeyer, J.; Grimm, L.; Hamilton, C.A.; Steiner, R.L.: *High precision measurement of a possible resistive slope of Josephson array voltage steps*, IEEE Electr. Dev. Lett. 7 (1986) 44–46

4.16 Kanter, H.; Vernon, F.L. *High-frequency response of Josephson point contacts*, J. Appl. Phys. 43 (1972) 3174–3183

4.17 Russer, P.: *Influence of microwave radiation on current-voltage characteristic of superconducting weak links*, J. Appl. Phys. 43 (1972) 2008–2010

4.18 Stephen, M.J.: *Noise in a driven Josephson oscillator*, Phys. Rev. 186 (1969) 393–397

4.19 Adde, R.; Vernet, G.: *High frequency properties and applications of Josephson junctions from microwaves to far-infrared*, Superconductor Applications: SQUIDs and Machines, Ed. B. B. Schwartz & S. Foner, New York: Plenum Press 1977, p. 249

4.20 Hartfuss, H.J.; Gundlach, K.H.; Schmidt, V.V.: *Nonhysteretic Josephson tunnel junctions for microwave detection*, J. Appl. Phys. 52 (1981) 5411–5413

4.21 Likharev, K.K.; Semenov, V.K.: *The characteristics of a superconducting point contact Josephson detector. Wide-band mode of operation*, Radio Eng. Electr. Phys. 18 (1975) 1734–1741

4.22 Kadlec, J.; Gundlach, K.H.; Hartfuss, H.J.: *The optimization of a Josephson video-detector for millimetre wavelength*, Infr. Phys. 19 (1979) 329–334

4.23 Likharev, K.K.; Semenov, V.K.: *The characteristics of a superconducting point contact Josephson detector. Selective mode of operation*, Radio Eng. Electr. Phys. 18 (1975) 1892–1899
4.24 Divin, Yu.Ya.; Polyanski, O.Yu.: *Incoherent radiation spectroscopy based on ac Josephson effect*, IEEE Trans. Mag. MAG-19 (1983) 613–615
4.25 Likharev, K.K.; Ulrich, B.T.: *Systems with Josephson junctions* (in Russian), Moscow State Univ. Press 1978
4.26 Divin, Yu.Ya.; Mordovets, N.A.: *Width of the Josephson-generation line in the far-IR region*, Sov. Tech. Phys. Lett. 9 (1983) 108–110
4.27 Likharev, K.K.; Semenov, V.K.: *Fluctuation spectrum in superconducting point junctions*, J. Electr. Phys. 15 (1972) 442–445
4.28 Rogovin, D.; Scalapino, D.J.: *Fluctuation phenomena in tunnel junctions*, Ann. Phys. 86 (1974) 1–90
4.29 Hubermann, B.A.; Crutchfield, J.P.; Packard, N.H.: *Noise phenomena in Josephson junctions*, Appl. Phys. Lett. 37 (1980) 750–752
4.30 Stumper, U.; Hinken, J.H.; Richter, W.; Schiel, D.; Grimm, L.: *Experimental investigation of a new spectrometer comprising a Josephson junction*, Electr. Lett. 20 (1984) 540
4.31 Bracewell, R.M.: *The Fourier transform and its applications*, New York etc.: McGraw-Hill 1965, p. 267
4.32 Taur, Y.: *Josephson-junction mixer analysis using frequency-conversion and noise-correlation matrices*, IEEE Trans. Electr. Dev. ED-27 (1980) 1921–1928
4.33 Taur, Y.; Claassen, J.H.; Richards, P.L.: *Conversion gain in a Josephson effect mixer*, Appl. Phys. Lett. 24 (1974) 101–103
4.34 Claassen, J.H.; Richards, P.L.: *Point-contact Josephson mixers at 130 GHz*, J. Appl. Phys. 49 (1978) 4130–4140
4.35 Taur, Y.; Kerr, A.R.: *Low-noise Josephson mixers at 115 GHz using a recyclable point contact*, Appl. Phys. Lett. 32 (1978) 775–777
4.36 Henaux, J.-C.; Vernet, G.; Adde, R.: *Far infrared frequency response of a Josephson junction in a self-pumped mixer*, J. Appl. Phys. 54 (1983) 7078–7082
4.37 Blaney, T.G.: *Josephson mixers of submillimetre wavelengths: Present experimental status and future developments*, New York: AIP Confer. Proc. 44 (1978), p. 230
4.38 McDonald, D.G.; Kose, V.E.; Evenson, K.M.; Wells, J.S.; Cupp, J.D.: *Harmonic generation and submillimeter wave mixing with the Josephson effect*, Appl. Phys. Lett. 15 (1969) 121–122
4.39 Russer, P.: *General energy relations for Josephson junctions*, Proc. IEEE 59 (1971) 282–283
4.40 Russer, P.: *Die Anwendung von Josephsonelementen in Mikrowellenempfängern (The application of Josephson junctions to microwave receivers)*, Nachrichtentechn. Zeitschr. 31 (1978) 604–612
4.41 Rudner, S.; Claeson, T.: *Advances in high frequency applications of superconducting tunnel junctions*, Superconducting Quantum Interference Devices and their Applications: SQUID '85 Ed. H.D. Hahlbolm & H. Lübbig. Berlin, New York: de Gruyter 1985, p. 963
4.42 Calander, N.; Claeson, T.; Rudner, S.: *Shunted Josephson tunnel junctions: High frequency, self-pumped low noise amplifiers*, J. Appl. Phys. 53 (1982) 5093–5103
4.43 Kuzmin, L.S.; Likharev, K.K.; Migulin, V.V.; Polunim, E.A.; Simonov, N.A.: *X-band parametric amplifier and microwave SQUID using single-tunnel-junctions superconducting interferometer*, Superconducting Quantum Interference Devices and their Applications: SQUID '85, Ed. H.D. Hahlbolm & H. Lübbig. Berlin, New York: de Gruyter 1985, p. 1029
4.44 Smith, A.D.; Sandell, R.D.; Burch, J.F.; Silver, A.H.: *Low noise microwave parametric amplifier*, IEEE Trans. Mag. MAG-21 (1985) 1022–1028
4.45 Russer, P.: *Ein gleichstromgepumpter Josephson-Wanderwellenverstärker (A d.c. pumped Josephson travelling wave amplifier)*, Wiss. Ber. AEG-Telefunken 50 (1977) 171–182
4.46 Rajeevakumar, T.V.: *A Josephson vortex-flow device*, Appl. Phys. Lett. 39 (1981) 439
4.47 Nagatsuma, T.; Enpuku, K.; Iwakura, H.; Yoshida, K.: *Flux-flow-type Josephson linear amplifier with large gain and wide linear range*, Jpn. J. Appl. Phys. 24 (1985) L599–L601
4.48 Yoshida, K.; Hashimoto, T.; Nagatsuma, T.; Enpuku, K.: *Josephson analog amplifier with large current gain*, IEEE Trans. Mag. MAG-23 (1987) 723–726

4.49 Hilbert, C.; Clarke, J.: *DC SQUIDs as radiofrequency amplifiers*, J. Low Temp. Phys. 61 (1985) 263–280

4.50 Van Duzer, T.; Turner, C.W.: *Principles of superconductive devices and circuits*, New York: Elsevier North Holland 1981, p. 204

4.51 Hogrefe, J.: *Subharmonisch injektionssynchronisierter Oszillator mit Josephsonelement (Subharmonic injection-synchronised oscillator with Josephson junction)*, Dipl.-Arbeit am Institut für Hochfrequenztechnik (Diploma-Thesis of the Institute of High Frequency Engineering) Technische Universität Braunschweig (Brunswick) 1985

4.52 Nagatsuma, T.; Enpuku, K.; Irie, F.: *Flux-flow type Josephson oscillator for millimeter and submillimeter wave region*, J. Appl. Phys. 54 (1983) 3302–3309

4.53 Nagatsuma, T.; Enpuku, K.; Yoshida, K.; Irie, F.: *Flux-flow Josephson oscillator for millimeter and submillimeter wave region. II. Modeling*, J. Appl. Phys. 56 (1984) 3284–3293

4.54 Nagatsuma, T.; Enpuku, K.; Sueoka, K.; Yoshida, K.; Irie, F.: *Flux-flow type Josephson oscillator for millimeter and submillimeter wave region. III. Oscillation stability*, J. Appl. Phys. 58 (1985) 441–449

4.55 Irie, F.; Yoshida, K.: *Problems of fluxon motion in long Josephson junctions and their applications*, Superconducting Quantum Interference Devices and their Applications: SQUID '85, Ed. H.D. Hahlbolm & H. Lübbig. Berlin, New York: de Gruyter 1985, p. 431

4.56 McColl, M.; Millea, M.F.; Silver, A.H.; Bottjer, M.F.; Pedersen, R.J.; Vernon, F.L. Jr.: *The Super-Schottky Microwave mixer*, IEEE Trans. Magnetics, MAG-13 (1977) 221–227

4.57 McColl, M.; Millea, M.F.; Silver, A.H.: *The superconductor-semiconductor Schottky barrier diode detector*, Appl. Phys. Lett. 23 (1973) 263

4.58 Kollberg, E.: *Superconducting vs Schottky mixers for mm and sub-mm waves*, in *Instrumentation for Submillimeter Spectroscopy*, Eric Kollberg, Editor, Proc. SPIE 598 (1986) 8–15

4.59 Lloyd, F.L.; Hamilton, C.A.; Beall, J.A.; Go, D.; Ono, R.H.; Harris, R.E.: *A Josephson Array Voltage Standard at 10 V*, IEEE Electr. Dev. Lett. EDL-8 (1987) 449–450

4.60 Hinken, J.H.: *The Josephson Junction as a Spectral Detector*, in "Superconducting Quantum Electronics", Ed. V. Kose, Berlin, Heidelberg, New York: Springer 1989

Chapter 5

5.1 Sze, S.M. (Ed.): *VLSI Technology*, Tokyo: McGraw-Hill 1983

5.2 Anderson, A.C.; Withers, R.S.; Reible, S.A.; Ralston, R.W.: *Substrates for superconductive analog processing devices*, IEEE Trans. Magnetics MAG-19 (1983) 485–489

5.3 Niemeyer, J.; Hinken, J.H.: *Präzise, mikrowelleninduzierte Normalspannungen von Reihenschaltungen mit mehr als 1000 Josephson-Tunnelelementen (Precise, microwave-induced standard voltages from series circuits with more than 1000 Josephson tunnel junctions)*, mikrowellenmagazin (Microwave Magazine) 13 (1987) No. 2, 118–124

5.4 IBM Journal of Research and Development, vol. 24, No. 2, March 1980, Special Issue on Josephson Computer Technology

5.5 Barone, A.; Paterno, G.: *Physics and Applications of the Josephson Effect*, New York etc.: Wiley 1982

5.6 Broom, R.F.; Jaggi, R.; Mohr, Th.O.; Oosenbrug, A.: *Effect of Process Variables on Electrical Properties of Pb-Alloy Josephson-Junctions*, [5.4], p. 206

5.7 Greiner, J.H.; Kircher, C.J.; Klepner, S.P.; Lahiri, S.K.; Warneke, A.J.; Basavaiah, S.; Yen, E.T.; Baker, J.M.; Brosious, P.R.; Huang, H.-C.W.; Murakami, M.; Ames, I.: *Fabrication Process for Josephson Integrated Circuits*, [5.4], p. 195

5.8 Gundlach, K.H.; Takado, S.; Zahn, M.; Hartfuss, H.J.: *A new lead alloy tunnel junction for quasiparticle mixer and other applications*, Appl. Phys. Lett. 41 (1982) 294–296

5.9 Imamaru, T.; Hoko, H.; Tamura, H.; Yoshida, A.; Suzuku, H.; Morohashi, S.; Ohara, S.; Hasuo, S.; Yamoaka, T.: *Fabrication technology for lead-alloy Josephson devices for high density integrated circuits*, J. Appl. Phys. 59 (1986) 1720–1748

5.10 Broom, R.F.; Laibowitz, R.B.; Mohr, Th.O.; Walter, W.: *Fabrication and Properties of Niobium Josephson Tunnel Junctions*, [5.4], p. 212

5.11 Beasley, M.R.; Kircher, C.J.: *Josephson junction electronics: materials issues and fabrication techniques*, in "Superconductor Material Science, Metallurgy, Fabrication and Applications", Eds.: Foner, S.; Schwarz, B.B.; New York: Plenum Press 1981, 605–684

5.12 Raider, S.I.; Drake, R.E.: unpublished results, according to [45] in [5.11]

5.13 Braginski, A.I.; Gavaler, J.R.; Jarocko, M.A.: *New Materials for refractory tunnel junctions: fundamental aspects*, in "Superconducting Quantum Interference Devices and their Applications", Eds: Hahlbolm, H.D.; Lübbig, H.; Berlin: Walter de Gruyter 1985, 591–629

5.14 Shoji, A.; Aayagi, M.; Kosaka, S.; Shinoki, F., Hayakawa, H.: *Niobium nitride Josephson tunnel junctions with magnesium oxide barriers*, Appl. Phys. Lett. 46 (1985) 1098-1100

5.15 Greiner, J.H.: *Oxidation of lead films by sputter etching in an oxygen plasma*, J. Appl. Phys. 45 (1974) 32–37

5.16 Morohashi, S.; Shinoki, F.; Shoji, A.; Hayakawa, H.: *High quality Nb/Al-AlOx/Nb Josephson junction*, Appl. Phys. Lett. 46 (1985) 1179–1181

5.17 Niemeyer, J.; Sakamoto, Y.; Vollmer, E.; Hinken, J.H.; Shoji, A.; Nakagawa, H.; Takado, S.; Kosaka, S.: *Nb/Al-oxide/Nb and NbN/MgO/NbN Tunnel Junctions in Large Series Arrays for Voltage Standards*, Jap. J. Appl. Phys. 25 (1986) L343–L345

5.18 Gundlach, K.H.; Blundell, R.; Blum, E.J.: *Eine Neuentwicklung für die Radioastronomie: Der SIS-Empfänger (New developments in radioastronomy: The SIS-receiver)*, mikrowellenmagazin (Microwave Magazine) 11 (1985) No. 1, 32–37

5.19 Ames, I.: *An Overview of Materials and Process Aspects of Josephson Integrated Circuit Fabrication*, [5.4], 188

5.20 Kroger, H.; Smith, L.N.; Jillie, D.W.: *Selective niobium anodization process for fabricating Josephson tunnel junctions*, Appl. Phys. Lett. 39 (1981) 280

5.21 Shoji, A.: *NbN based Josephson junctions*, in "Superconducting Quantum Interference Devices and their Applications", Eds.: Hahlbolm, H.D.; Lübbig, H.; Berlin: Walter de Gruyter 1985, 631–657

5.22 Shoji, A.; Shinoki, F.; Kosaka, S.; Aayagi, M.; Hayakawa, H.: *New fabrication process for Josephson tunnel junctions with (niobium nitride, niobium) double-layered electrodes*, Appl. Phys. Lett. 41 (1982) 1097

5.23 Gurvitch, M.; Washington, M.A.; Huggins, H.A.: *High quality refractory Josephson tunnel junctions utilizing thin aluminium layers*, Appl. Phys. Lett. 42 (1983) 472

5.24 Mück, M.; Rogalla, H.; David, B.; Heiden, C.: *Response of Nb_3Ge Microbridges to Microwave Irradiation*, Zeitschr. Phys. B-Condensed Matter, 61 (1985) 81–84

5.25 van Dover, R.B.; Howard, R.E.; Beasley, M.R.: *Fabrication and characterization of S-N-S planar microbridges*, IEEE Trans. Magn. MAG-15 (1979) 574

5.26 Zimmermann, J.E.; Thiene, P.; Harding, J.T.: *Design and operation of stable rf-biased superconducting point-contact quantum devices and a note on the properties of perfectly clean metal contacts*, J. Appl. Phys. 41 (1970) 1572–1580

5.27 Edrich, J.: *A Low-Noise 47-GHz Mixer Using a Permanent Josephson Junction*, IEEE Trans. Microwave Theory Techn., MTT-24 (1976) 706–709

5.28 Vowinkel, B.: *Untersuchungen an Josephson-Punktkontakt-Mischern im Millimeterwellen-Bereich (Investigations into Josephson point contact mixers in the millimetre wave region)*, Nachrichtentechn. Zeitschr. Archiv 2 (1980) 151–154

5.29 Taur, Y.; Kerr, A.R.: *Low-noise Josephson mixers at 115 GHz using recyclable point contacts*, Appl. Phys. Lett. 32 (1978) 775–777

5.30 Zimmermann, J.E.: *A review of the properties and applications of superconducting point contacts*, Proc. Applied Superconductivity Conference 1972, 544–561

5.31 Bednorz, J.G.; Müller, K.A.: *Possible High T_c Superconductivity in the Ba-La-Cu-O System*, Zeitschr. Phys. B, Condensed Matter 64 (1986) 189–193

5.32 Ihara, H.; Terada, N.; Masatashi, Jo; Hirahayashi, M.; Tokumato, M.; Kumura, Y.; Matsubara, T.; Sugise, R.: *Possibility of Superconductivity at 65° C in Sr-Ba-Y-Cu-O System*, Jap. J. Appl. Phys. 26 (1987)

5.33 Wu, M.K.; Ashburn, R.J.; Torng, C.J.; Hor, P.H.; Meng, R.L.; Gao, L.; Huang, Z.J.; Wang, Y.Q.; Chu, C.W.: *Superconductivity at 93 K in a New Mixed-Phase Y-Ba-Cu-O Compound System at Ambient Pressure*, Phys. Rev. Lett. 58 (1987) 908–910

5.34 Maeda, H.; Tanaka, Y.; Fukutomi, M.; Asano, T.: *A New High-T_c Oxide Superconductor without a Rare Earth Element*, Jap. J. Appl. Phys. Lett. 27 (1988) L209–L210

5.35 Green, S.M.; Jiang, C.; Mei, Yu.; Luo, H.L.; Politis, C.: *Zero Resistance at 107 K on the (Bi, Pb)-Ca-Sr-Cu-Oxide System*, Phys. Rev. B 38 (1988) 5016–5018

5.36 Parkin, S.S.P.; Lee, V.Y.; Engler, E.M.; Nazzal, A.I.; Huang, T.C.; Gorman, G.; Savoy, R.; Beyers, R.: *Bulk Superconductivity at 125 K in* $Tl_2Ca_2Ba_2Cu_3O_x$, Phys. Rev. Lett. 60 (1988) 2539–2542

5.37 Gopolakrishnan, I.K.; Sastry, P.V.P.S.S.; Gangadharan, K.; Phatak, G.M.; Yakhmi, J.V.; Iyer, R.M.: *Synthesis and Properties of a 125 K Superconductor in the Tl-Ca-Ba-Cu-O System*, Appl. Phys. Lett. 53 (1988) 414–416

5.38 Beyers, R.; Lim, G.; Engler, E.M.; Lee, V.Y.; Ramirez, M.L.; Savoy, R.J.; Jacowitz, R.D.; Shaw, T.M.; LaPlaca, S.; Boehme, R.; Tsuei, C.C.; Park, S.I.; Shafer, W.W.; Gallagher, W.J.; Chandrashekhar, G.V.: *Annealing Treatment Effects on Structure and Superconductivity in* $Y_1Ba_2Cu_3O_{9-x}$, Appl. Phys. Lett. 51 (1987) 614–616

5.39 Crabtree, G.W.; Kwok, W.K.; Umezawa, A.: in Weber, H.W. [ed.] *High-Tc Superconductors*, New York: Plenum Press 1988 p. 233, cited as [41] in [5.40]

5.40 Weber, H.W.: *Magnetization of Single Crystalline and Grain Aligned High-Tc Superconductors*, in Narlikar, A.V. [ed.]: *Studies of High-Temperature Superconductors*, Commack: Nova Science 1989

5.41 Bezinge, A.; Jorda, J.L.; Junod, A.; Muller, J.: *Magnetization of the Extreme Type II Superconductor* $YBa_2Cu_3O_7$ *with* $\kappa > 100$, Solid State Comm. 64 (1987) 79–82

5.42 Malozemoff, A.P.; Grand, P.M.: *High Temperature Superconductivity Research at IBM Thomas J. Watson and Almaden Research Centers*, Zeitschr. Phys. B. Condensed Matter 76 (1987) 275–283

5.43 Crommie, M.F.; Bourne, L.C.; Zettl, A.; Cohen, M.L.; Stacy, A.: *Tunneling measurement of the energy gap in Y-Ba-Cu-O*, Phys. Rev. B 35 (1987) 8853–8855

5.44 Junod, A.; Bezinge, A.; Graf, T.; Jorda, J.L.; Muller, J.; Antognazza, L.; Cattani, D.; Cors, J.; Decraux, M.; Fischer, O.; Banovski, M.; Genoud, P.; Hoffmann, L.; Manuel, A.A.; Peter, M.; Walker, E.; Francois, M.; Yvon, K.: *Structure, Resistivity, Critical Field, Specific-Heat Jump at* T_c*, Meissner Effect, a.c. and d.c. Susceptibility of the High-*T_c *Superconductor* $YBa_2Cu_3O_7$, Europhys. Lett. 4 (1987) 247–252

5.45 Eickenbusch, H.; Paulus, W.; Schöllhorn, R.; Schlögl, R.: *Bulk and surface characteristics of the single phase high* T_c *superconductor* $Y_{11}Ba_{213}CuO_{3-n}$, Mat. Res. Bull. 22 (1987)

5.46 Halbritter, J.: *Percolation in Superconducting Cuprates: Resistivity and Critical Currents*, Int. J. Mod. Phys. B. (1989)

5.47 Kawasaki, M.; Nagata, S.; Sato, Y.; Funabashi, R.; Hasegawa, T.; Kishio, K.; Kitazawa, K.; Fueli, K.; Koinuma, H.: *High* T_c *Yb-Ba-Cu-O Thin Films Deposited on Sintered YSZ Substrates by Sputtering*, Jap. J. Appl. Phys. 26 (1987) L738–L740

5.48 Itozaki, H.; Tanaka, S.; Harada, K.; Fujimori, N.; Yazu, S.: *Properties of High* T_c *Y-Ba-Cu-O Thin Films Prepared by Sputtering*, Ext. Abstr. 1987 Int. Supercond. Electronic Conf. (ISEC '87), Tokyo, 407

5.49 Chaudhari, P.; Koch, R.H.; Laibowitz, R.B.; McGuire, T.R.; Gambino, R.J.: *Critical-Current Measurements in Epitaxial Films of* $YBa_2Cu_3O_{7-x}$ *Compound*, Phys. Rev. Lett 58 (1987) 2684–2686

5.50 Dijkkamp, D.; Venkatesan, T.; Wu, X.D.; Shaheen, S.A.; Jisrawi, N.; Min-Lee, Y.H.; McLean, W.L.; Croft, M.: *Preparation of Y-Ba-Cu oxide superconductor thin films using pulsed laser evaporation from high* T_c *bulk material*, Appl. Phys. Lett. 51 (1987) 619–621

5.51 Kwo, J.; Hong, M.; Trevor, D.J.; Fleming, R.M.; White, A.E.; Farrow, R.C.; Kortan, A.R.; Short, K.T.: *In situ Epitaxial Growth of* $Y_1Ba_2Cu_3O_{7-x}$ *Films by Molecular Beam Epitaxy with an Activated Oxygen Source*, Appl. Phys. Lett. 53 (1988) 2683–2685

5.52 *Bibliography of High-Tc Superconducting Films*, prepared by J. Talvaccio, January 1989, Westinghouse R&D Center, Pittsburgh, Pennsylvania, and Lawrence Berkeley Laboratory, University of California, LBL-26578, UC-404, available from National Technical Information Service, U.S. Department of Commerce, 5285 Port Royal Road, Springfield, VA 22161 U.S.A.

5.53 Beasley, M.R.: *Advances in the Fabrication of High-Temperature Superconductors*, Physics Today, January 1989, 23–24

5.54 Lathrop, D.K.L.; Russek, S.E.; Buhrmann, R.A.: *Production of $YBa_2Cu_3O_{7-y}$ Superconducting Thin Films in situ by High-Pressure Reactive Evaporation and Rapid Thermal Annealing*, Appl. Phys. Lett. 51 (1987) 1554–1556

5.55 Terashima, T.; Iijima, K.; Yamamoto, K.; Bando, Y.; Mazako, H.: *Single-Crystal $YBa_2Cu_3O_{7-x}$ Thin Films by Activated Reactive Evaporation*, Jpn. J. Appl. Phys. 27 (1988) L91–L93

5.56 Silver, R.M.; Berezin, A.B.; Wendmann, W.; de Lozanne, A.L.: *As-Deposited Superconducting Y-Ba-Cu-O Thin Films on Si, Al_2O_3 and $SrTiO_3$ Substrates*, Appl. Phys. Lett. 52, (1988) 2174–2176

5.57 Spah, R.J.; Hess, H.F.; Stormer, H.L.; White, A.E.; Short, K.T.: *Parameters for in situ Growth of High Tc Superconducting Thin Films Using an Oxygen Plasma Source*, Appl. Phys. Lett. 53 (1988) 441–443

5.58 Berberich, P.; Tate, J.; Dietsche, W.; Kinder, H.: *Low-Temperature Preparation of Superconducting $YBa_2Cu_3O_{7-\delta}$ Films on Si, MgO, and $SrTiO_3$ by Thermal Coevaporation*, Appl. Phys. Lett. 53 (1988) 925–926

5.59 Wasa, K.; Kitabatake, M.; Adachi, H.; Setsune, K.; Hirochi, K.: in Harper, J.; Cotton, R.; Feldmann, L. (eds.): *Thin Film Processing and Characterization of High-Temperature Superconductors*, AIP Conf. Proc. No. 165 (AIP, New York, 1988), 38

5.60 Lin, R.J.; Chen, Y.C.; Kung, J.H.; Wu, P.T.: in Brodsky, M.; Dynes, R.; Kitazaw, K.; Tuller, H.: *Materials Research Society Symposium Proceedings*, (MRS, Pittsburgh, 1988), Vol. 99, 319

5.61 Dijkkamp, D.; Venkatesan, T.; Wu, X.D.; Shakeen, S.A.; Jirasani, N.; Min-Lu, Y.M.; McLean, W.L.; Croft, M.: *Preparation of Y-Ba-Cu Oxide Superconductor Thin Films Using Pulsed Laser Evaporation from High Tc Bulk Material*, Appl. Phys. Lett. 51 (1987) 619–621

5.62 Roas, B.; Schultz, L.; Endres, G.: *Epitaxial Growth of $YBa_2Cu_3O_{7-x}$ Thin Films by Laser Evaporation Process*, Appl. Phys. Lett. 53 (1988) 1557–1559

5.63 Klein, N.; Müller, G.; Piel, H.; Roas, B.; Schultz, L.; Klein, U.; Peiniger, M.: *Millimeter Wave Surface Resistance of Epitaxially Grown $YBa_2Cu_3O_{7-x}$ Thin Films*, Appl. Phys. Lett. 54 (1989) 757–759

5.64 Daginnus, M.; Hinken, J.H.: *A 77 K Computer Controlled mm-Wave Spectral Josephson Detector Using Microstrip Technique, Preliminary Results*. To be published in IEEE Microw. Theory. Techn. Symp. Digest, June 1989, Long Beach California

5.65 Häuser, B.; Klopman, B.B.G.; Gerritsma, G.J.; Gao, J.; Rogalla, H.: *Response of YBaCuO Thin Film Microbridges to Microwave Irradiation*, Appl. Phys. Lett. 54 (1989) 1368

5.66 Zimmermann, J.E.; Beall, J.A.; Cromar, M.W.; Ono, R.H.: *Operation of a Y-Ba-Cu-O rf SQUID at 81 K*, Appl. Phys. Lett. 51 (1987) 617–618

5.67 Zavaritzky, N.V.; Zavaritsky, V.N.: *High-Tc Ceramic Weak Links, rf-SQUIDs and their Applications*, to be published in Progress in High Temperature Superconductivity, Vol. 11: "High Tc from Russia", World Scientific Publ.

5.68 Bobrakov, V.F.; Vasiliev, B.V.; Polushkin, V.N.: *SQUID Operating at Liquid Nitrogen Temperatures*, JINR Rapid Comm. 30 (1988) 101–106

5.69 Gough, C.E.: *Flux Quantisation and SQUID Magnetometry Using Ceramic Superconductors*, Proc. Int. Conf. on High Temperature Superconductors and Materials and Mechanisms of Superconductivity, Interlaken, 1988, Part II, 1569–1573, see also Physica C 153–155 (1988) 1569–1573

5.70 Koch, R.H.; Umbach, C.P.; Clark, G.J., Chaudhari, P.; Laibowitz, R.B.: *Quantum interference devices made from superconducting oxide thin films*, Appl. Phys. Lett. 51 (1987) 200–202

5.71 Pegrum, C.M.; Donaldson, G.B.: *Probable limits of high T_c DC SQUID sensitivity*, Ext. Abstr. 1987 Int. Supercond. Electronics Conf. (ISEC '87), Tokyo, 405

5.72 Schwartz, D.B.; Mankiewich, P.M.; Howard, R.G.; Dayem, A.H.; Jackel, L.D.; Burkhardt, E.G.; Straughn, B.L.: *The Observation of AC Josephson Effect in a* $YBa_2Cu_3O_7/Au/YBa_2Cu_3O_7$ *Junction*, IEEE Transactions Mag. MAG-25 (1989) 1298–1300

5.73 Shiota, T.; Takechi, K.; Takai, Y.; Hayakawa, H.: *An Observation of Quasi-Particle Tunneling Characteristics in all Y-Ba-Cu-O Thin Film Tunnel Junction*, Proc. First Intern. Symp. on Superconductivity, Nagoya, 1988

Chapter 6

6.1 Frey, H.; Haefer, R.A.: *Tieftemperaturtechnologie (Low temperature technology)*, Düsseldorf: VDI-Verlag 1981

6.2 Barron, R.F.: *Cryogenic Systems*, New York: Oxford University Press 1985

6.3 Fastowski, W.G.; Petrowski, J.W.; Rowinski, A.E.: *Kryotechnik (Cryogenic technology)*, Berlin: Akademie-Verlag 1970

6.4 Klipping, G.: *Kryotechnik (Cryogenic technology)*, in "Handbuch Supraleitertechnik" ("Handbook of Superconductor Technology") at the seminar of the same name, Düsseldorf: VDI-Bildungswerk 1981, BW 2789

6.5 Hansen, H.; Linde, H.: *Tieftemperaturtechnik (Low Temperature Technique)*, 2nd ed. Berlin etc.: Springer 1985

6.6 Vowinkel, B.: *Untersuchungen an Josephson-Punktkontakt-Mischern im Millimeterwellen-Bereich (Investigations into Josephson point contact mixers in the millimetre wave region)*, Nachrichtentechn. Zeitschr. Archiv 2 (1980) 151–154

6.7 Blundell, R.; Hein, H.; Gundlach, K.H.; Blum, E.J.: *An SIS-Receiver for the 3mm Wavelength Range*, Int. J. Infrared Millimeter Waves 3 (1982) 793–799

6.8 Hilberath, W.; Vowinkel, B.: *Closed cycle refrigerator for superconducting mm wave-mixers*, Cryogenics 25 (1985) 573–577

Subject Index